Analog-Digital
and
Digital-Analog
Conversion

Analog-Digital and Digital-Analog Conversion

BERNARD LORIFERNE

Centre Technique de l'Industrie Horlogère,
Besançon, France

 LONDON · PHILADELPHIA · RHEINE

Heyden & Son Ltd, Spectrum House, Hillview Gardens, London NW4 2JQ, UK
Heyden & Son Inc, South 41st Street, Philadelphia, PA 19104, USA
Heyden & Son GmbH, Devesburgstrasse 6, 4440 Rheine, West Germany

---

British Library Cataloguing in Publication Data

Loriferne, Bernard
  Analog–digital and digital–analog conversion.
  1. Digital-to-analog converters
  2. Analog-to-digital converters
  I. Title   II. La conversion analogique-numérique, numérique-analogique. *English*
  621.3819′596   TK7887.6
  ISBN 0-85501-497-0

Title of French original edition:
La Conversion Analogique–Numérique, Numérique–Analogique
Compagnie Française d'Editions, Paris, 1976
© Les Editions de l'Usine Nouvelle, Paris, 1981
ISBN 2-7327-0016-9

Translation by:
Middlesex Polytechnic Consultancy Services Organization

Printed in Great Britain by J. W. Arrowsmith Ltd
Bound in Great Britain by Mackays of Chatham Ltd

CONTENTS

Preface   ix
    Wherefore Conversion?   ix
    Analog Signals and Digital Signals   ix

1   Basic Principles of A $\rightleftharpoons$ D Conversion   1
  1.1   Retrospect   1
  1.2   Fields of Application   3
      1.2.1   Instrumentation and Automatic Testing   3
      1.2.2   Arithmetical and Trigonometrical Operations   4
      1.2.3   Communications and Signal Analysis   5
      1.2.4   Visual Displays   6
      1.2.5   Industrial   6
  1.3   Operations Carried out During Conversion   7
  1.4   Theoretical Considerations   9
      1.4.1   Sampling   9
      1.4.2   Recovery   12
      1.4.3   Quantization   14
  1.5   Codes   17
      1.5.1   Unipolar Codes   18
      1.5.2   Bipolar Codes   20
  1.6   Sample and Hold Circuits   22
      1.6.1   The Operation of an Ideal Sampling Device   23
      1.6.2   Errors in Sample and Hold Circuits   25
          1.6.2.1   Sample   25
          1.6.2.2   The transition from sample to hold   25
          1.6.2.3   Hold   26
          1.6.2.4   The transition from hold to sample   27
      1.6.3   Some Examples of Sample Circuits   28

2   Digital to Analog Conversion   30
  2.1   Definition   30
  2.2   The Different Types of DAC   32
  2.3   Characteristic Parameters of a DAC   33
  2.4   Errors in DACs   38
      2.4.1   Offset Error   38
      2.4.2   Gain Error   39
      2.4.3   Linearity Error   39

		2.4.4	Differential Linearity—Monotonicity	40
		2.4.5	The Influence of Temperature	42
	2.5	Analysis of the Main DACs		44
		2.5.1	Parallel DACs	44
			2.5.1.1 Weighted resistor converter	45
			2.5.1.2 The ladder converter	47
			2.5.1.3 DACs using bipolar codes	49
			2.5.1.4 Example of a commercial DAC	51
			2.5.1.5 Multiplier DACs	53
			2.5.1.6 The advantages of an input interface	54
		2.5.2	Serial DACs	55
		2.5.3	Indirect Converters	58
			2.5.3.1 The intermediate parameter DAC using pulses	58
			2.5.3.2 Stochastic converters	60
	2.6	Measurements of the Characteristics of a DAC		62
		2.6.1	The Measurement of Linearity	62
		2.6.2	The Measurement of the Conversion Time	64
	2.7	Examples		65
3	**Analog to Digital Conversion**			**67**
	3.1	Definition		67
	3.2	Classification of ADCs		68
	3.3	Characteristic Parameters of an ADC		69
		3.3.1	Ideal Transfer Function of an ADC	69
		3.3.2	Resolution	70
		3.3.3	Conversion Time	70
		3.3.4	Accuracy	71
		3.3.5	Noise Rejection	71
		3.3.6	Bipolar ADCs	72
	3.4	Errors in ADCs		72
		3.4.1	Quantization Error	73
		3.4.2	Offset Error	73
		3.4.3	Gain Error	74
		3.4.4	Linearity Error	74
		3.4.5	Differential Linearity	76
		3.4.6	Temperature Effects	77
	3.5	Analog ADCs		77
		3.5.1	Single-slope Converters	77
		3.5.2	Dual-slope Converters	79
		3.5.3	Counter Ramp Converters	82
		3.5.4	Continuous Counter Ramp Converters	84
		3.5.5	Voltage to Frequency Converters	85
		3.5.6	Capacitive Charge Transfer Converters	88
	3.6	Logic ADCs		92
		3.6.1	Parallel ADCs	92
		3.6.2	Successive Approximation Converters	95

	3.6.3	Organization and Operation	98
	3.6.4	Performance	101
3.7	Very High Speed ADCs		102
3.8	Bipolar ADCs		105
3.9	The Effects of Noise on A–D Conversion		106
	3.9.1	Noise in the Comparator	107
	3.9.2	Noise factor of a Converter	108
	3.9.3	A Brief Note on Noise	109
	3.9.4	Encoding Errors Introduced by Noise	110
3.10	Criteria for Testing an ADC		113
	3.10.1	Calibration	113
	3.10.2	Testing an ADC	114
	3.10.3	Statistical Test Methods	116
3.11	Examples		119
3.12	Logarithmic Converters		121
	3.12.1	LADCs Using an Analog Logarithmic Converter	122
		3.12.1.1 Logarithmic amplifiers	122
		3.12.1.2 Linear approximation by parts	123
		3.12.1.3 Discharge of an RC circuit	123
	3.12.2	LADCs Using a Digital Logarithmic Converter	123
		3.12.2.1 Sequential floating point converters	124
		3.12.2.2 Combining floating point converters	125
		3.12.2.3 Incremental logarithmic converters	125
	3.12.3	LADCs Using Non-linear DACs	126
		3.12.3.1 Weighted resistor exponential DACs	127
		3.12.3.2 Exponential DACs using ladder networks	128
		3.12.3.3 LADCs using an exponential DAC	129

4	**Digital–Synchro and Synchro–Digital Converters**		**131**
4.1	Angular Sensors		131
	4.1.1	Disc Encoders	131
		4.1.1.1 Incremental encoders	131
		4.1.1.2 Absolute encoders	132
	4.1.2	Synchros and Resolvers	133
	4.1.3	Comparison	135
4.2	Digital–AC Converters		135
	4.2.1	DAlC Characteristics	136
	4.2.2	DAlCs which use a Parallel DAC	136
	4.2.3	DAlCs for DC–AC Conversion	137
	4.2.4	DAlCs Using a Transformer	138
	4.2.5	Digital Synchro Converters	139
		4.2.5.1 Sin θ and cos θ available in digital form	140
		4.2.5.2 Sin θ and cos θ available in analog form	141
		4.2.5.3 Obtaining sin θ and cos θ	141
4.3	AC–Digital Conversion		144
	4.3.1	General	144

4.3.2 AC–DC Conversion 145
4.3.3 Single Phase Converters 147
4.3.4 Synchro–Digital Converters 148
4.3.5 SDCs giving $\sin\theta$ and $\cos\theta$ in Digital Form 149
4.3.6 SDCs giving θ in Digital Form 149
    4.3.6.1 Octant selector 150
    4.3.6.2 Tracking converter 151
    4.3.6.3 Successive approximation SDCs 154
    4.3.6.4 SDCs using ROMs 155
4.4 Examples 156

5 Components Used in Converters 158
5.1 Comparators 158
  5.1.1 Definition of a Comparator 158
  5.1.2 The Difference Between an Operational Amplifier
      and a Comparator 159
  5.1.3 Types of Comparators 160
  5.1.4 Comparator's Characteristics 161
    5.1.4.1 Ideal transfer characteristic 161
    5.1.4.2 Accuracy 162
    5.1.4.3 Speed 162
    5.1.4.4 Stability 163
    5.1.4.5 Technology of production 164
5.2 Resistance Networks 165
  5.2.1 Resistors Obtained by Diffusion 166
  5.2.2 Use of Ionic Implantation 167
  5.2.3 Film Resistors 168
  5.2.4 Comparison of the Methods 169
5.3 Analog Switches 170
  5.3.1 General 170
  5.3.2 Bipolar Transistor Switches 172
  5.3.3 FETs as Switches 173
    5.3.3.1 Junction gate FET 173
    5.3.3.2 Insulated gate FET (MOS) 174
    5.3.3.3 Comparison 174
  5.3.4 Junction FET Switches 175
  5.3.5 CMOS Switches 176
  5.3.6 Examples of Actual Devices 178
5.4 Reference Sources 181
  5.4.1 Zener Diode Reference 182
  5.4.2 Reference Using Bipolar Transistors 186

Bibliography 189

Index 191

PREFACE

WHEREFORE CONVERSION?

It is becoming common in many fields to contrast or compare analog and digital quantities. Thus, in the field of measurements, digital indicating instruments are being increasingly used to display readings. Analog data transmission and digital data transmission are competing in the field of communications etc., but the very existence of this competition is linked to the possibility of passing from an analog signal to its digital equivalent or vice versa, and these operations assume converters to be available.

The first chapter of the book is an introduction to present day analog $\rightleftharpoons$ digital conversion techniques. It should help towards a better understanding of the operation of converters and the problems that could be encountered when they are inserted in a system. The various operations which must be carried out when it is required to convert an analog quantity into its equivalent digital quantity and vice versa (sampling, quantization, filtering, etc.) are defined, and their essential characteristics are given. The few mathematical techniques and derivations used for that purpose should not deter the reader who is not too familiar with the mathematics involved as the following chapters will be devoted mainly to aspects of conversion techniques, highlighting problems related to their use.

ANALOG SIGNALS AND DIGITAL SIGNALS

Before examining the various analog to digital (A–D) and digital to analog (D–A) conversion processes it is useful to review the properties of each type of representation; in particular this may help select the representation most suited to the problem at hand.

An *analog signal* is a signal whose value varies continuously with time, its instantaneous amplitude itself varying continuously within a limited range. The simplest example is that of a sinusoidal signal $A \sin(\omega t + \phi)$, whose instantaneous value covers all the values within the range $(-A, +A)$. An analog signal may very often be expressed as a weighted sum of sinusoidal signals.

The analog signal is a simple type of signal which can quite conveniently be transmitted, but its simplicity results in several drawbacks. It is sensitive to parasitic signals and its amplitude or phase can be distorted by the transmitting system. When it undergoes such operations as analog multiplication the accuracy with which the signal is known is often reduced. Moreover it is difficult to store an analog signal.

On the other hand a *digital signal* usually appears as a series of symbols. Thus, in a binary system a signal consists of a series of numbers, each of which is 0 or 1, that may be given physical form by the absence or presence of pulses. It can be said that the signal is represented by a word of a given format that is of a given structure. The digital signal represents the value of a quantity at a specific instant. It is not a continuous signal and since the symbols making up this signal (usually numbers) can only vary by 'steps' the value represented by the signal must, perforce, be discrete.

These differences from the analog signal, which may appear to be obstacles, are largely compensated by the advantages gained by the digital representation. First of all a digital signal is far less sensitive to the imperfections of the transmitting system (distortion, noise) since it is only necessary to detect the pulses in order to obtain the information, their precise characteristics (amplitude, duration) not being taken into account. During the various operations the accuracy of the signal is maintained (if some of the truncating that is sometimes carried out is ignored). On the other hand the pass band required for transmitting digital information is much greater than the pass band needed for an analog representation of the information.

Thus, each representation has its advantages and disadvantages; the choice of one or other method must take into account the nature of the signal available, the possibility of introducing a digital process, the likelihood of transmitting the information, etc.

1 BASIC PRINCIPLES OF A⇌D CONVERSION

What is the present state of A⇌D conversion technology? Which new field of application will it now challenge? What are the advantages and disadvantages of stochastic conversion? What impact will complementary metal oxide semiconductor (CMOS) technology have on manufacturing techniques and on cutting costs? These are some of the topical questions that are bound to be asked. A quick glance through the technical literature will show that there is hardly a journal that does not have an article or carry an advertisement concerning A⇌D conversion and such a vast mass of information would in itself be ample justification for this book, devoted to the various methods of A⇌D conversion. Another no less important reason is the ever increasing use of converters; each day new areas are taken over by these techniques. It therefore seemed appropriate to us to undertake as exhaustive a survey as possible of existing converters, explaining their operation, highlighting their performance and giving an indication of the very many likely fields of application. This approach should help the user (who is not necessarily a specialist) choose from the manufacturers' literature the converter most suited to their requirements with the best trade off between quality and cost.

1.1 RETROSPECT

Our world is essentially a world in which information is available in analog form; every parameter which is to be investigated (speed, temperature, etc.) varies in a continuous way; therefore man's first involvement was with analog-type signals. Since the detector used produced analog signals and the control of machines or processes often required signals of the same type, it seemed logical to have adopted the view that signals should be handled in analog form. The increasing importance of digital signals compels us to alter this point of view. Although digital signals are more easily stored or transmitted and lend themselves better to computation than their analog counterparts, their applications only increased dramatically thanks, largely, to the opportunities opened up by digital computers and to the substantial reduction in converter costs brought about by technological advances and booming demand (in fact converters are used in many interfacing applications). The advent of monolithic technology for the fabrication of semi-conductor devices has brought about great advances in the field of $A \rightleftharpoons D$ converters, as in all areas utilizing electronic components, and its impact will continue to be felt. Monolithic technology profoundly influences the design concepts, the fabrication processes and the utilization of the devices for con-verting analog data into digital data or vice versa. These new techniques have been brought to bear on traditional products which were often conceived and designed for high accuracy instrumentation applications or for computations and are also themselves the 'breeding ground' for an entirely new range of products; thus, only the advent of large-scale integration technology (LSI) made it possible to produce random variable $A \rightleftharpoons D$ converters at competitive prices. Twenty years ago an analog to digital converter of 0.1% accuracy (i.e. supplying digital data in 10 bit form) and with a repetition rate of 50 kHz (that is, effecting one conversion every 20 µs) would have cost about £5000 including power supplies and control circuits. It would have been a valve-type converter, dissipating 500 W and occupying a volume of several dm³. A converter which has the same characteristics will now dissipate less than 1 W, occupy a volume of a few mm³ and cost less than £100. Thanks to CMOS technology it is possible to manufacture 10 bit digital to analog converters having a power dissipation of 20 mW! These technological advances have brought about improvements in the characteristics obtainable and at the same time have helped to reduce the size and power dissipation of the devices. Thus manufacturers are now producing 16 bit analog to digital converters with a self-calibrating facility. Moreover, these converters are generating an ever increasing field of applications. Twenty years ago converters were built up from a number of discrete components, with matched valves or transistors and manually trimmed wound resistors. Nowadays manufacturers use hybrid or LSI technologies. Advanced integration using MOS technology results in high working speeds with good accuracy since the relative drift between components is small. Monolithic 12 bit digital to analog or analog to digital converters are already being produced in dual-in-line (DIL) packages, and hybrid technology is still used when greater accuracy is required. Technological progress is an irreversible and continual process (ionic implantation etc.) but its impact on the design and production of $A \rightleftharpoons D$ converters depends on the price reductions that might result and on the potential markets for modules thus manufactured, for the financial profitability of the design must be ascertained before it can be undertaken. The converter market is very closely linked to the state of the art of sensors so that one may wonder why it is that no sensor has been manufactured which gives some kind of digital output, thus doing away with the analog to digital converter. This problem has been under investigation for a long time and no definite solution has yet been formulated. It is possible to manufacture digital sensors suitable for certain physical quantities but for other quantities it still remains a very difficult problem. Moreover, these sensors remain specialized devices and it is only with great difficulty that the section which carries out the conversion process can be used effectively elsewhere. Thus, the problem has not been solved but transferred from the converter to the sensor. Taking into consideration already existing converters, it seems more logical to retain analog sensors and add on non-specialized converters. The only foreseeable development will perhaps be to make the converter an integral part of the sensor; the space required to accommodate the converter should gradually become smaller as new techniques are introduced.

1.2 FIELDS OF APPLICATION

It would seem, perhaps, ambitious and even unrealistic to attempt to draw up an exhaustive list of all likely converter applications, so our purposely limited objective is to group the main applications into categories and give some examples of each category (Sections 1.2.1 to 1.2.5).

1.2.1 Instrumentation and Automatic Testing

The use of converters will trigger the design of new equipment (function generators or programmable DC sources) and bring about the automation of test benches by linking them to computers.[1]

Voltage sources. A properly calibrated digital–analog converter (DAC) is probably the simplest accurate voltage source obtainable. When operated manually it is used as an adjustable reference source and when computer controlled it can be used within a test bench facility. It is often necessary to follow the DAC by a power amplification stage in order to obtain the required current; the value of this current must be carefully chosen.

Current sources. The output signal of a DAC can be a current output and it can then be used as a current source with such advantages as low costs, simplicity, possibility of earthing the load, etc. In normal current sources the reference value is obtained by means of a precision potentiometer or a Zener diode but by using a DAC a programmable current source can easily be obtained; in that mode of operation the DAC output is usually at zero potential and it is necessary to provide an 'adapter' circuit to allow current to be drawn off.

Signal generators. A number which increases linearly with time can be obtained using clocks and counters. If this number is used as the input to a DAC then the output is a signal which varies linearly and whose amplitude increases in steps. If a read only memory (ROM) is inserted between the counter and the DAC then the law governing the variation with time of the number given by the counter can be altered and under these conditions the output signal from the DAC can have any shape whatever. All sorts of signal can therefore be generated by this method. If the input signal to a DAC is from a standard binary counter then the output from the DAC will be a saw tooth waveform; for each pulse, the counter is incremented by unity and the output signal of the DAC is increased by a certain amount. If an up and down counter is used then a triangular waveform is obtained; when the counter reaches its maximum capacity instead of going back to zero as in the case of the saw tooth generator, it starts counting down thus resulting in a constant reduction of the output signal of the DAC. If the counter drives a programmed ROM to provide a digitized sinusoidal signal and if it is followed by a DAC capable of reproducing the amplitude and sign of the signal, then a sinusoidal signal is obtained whose frequency depends upon that of the clock and whose amplitude can be controlled externally by varying the reference voltage feeding the DAC.

Digital voltmeters. It is not possible to discuss the applications of analog to digital converters (ADCs) in instrumentation without mentioning digital voltmeters; the ADC is the basic component of any digital voltmeter. A digital voltmeter must measure a voltage and display its value as a number of digits usually in decimal notation. An ADC will carry out precisely this operation using as a rule a binary code; to obtain a digital voltmeter it is only necessary to follow the ADC with a decoder/encoder converting the binary into decimal. In a digital voltmeter the output of the ADC is often available and can be connected directly to a computer (without code changing) so that the digital voltmeter can be used as an interface between a sensor and a computer.

An ADC can also be used in conjunction with any analog sensor and in this way the magnitude of the physical quantity being investigated will be available in digital form. The flexibility of the method allows either the value of the magnitude or simply its variations to be made available; the method can then be used to adjust the temperature of an oven, the travel of a machine, etc.

Automatic testing. Automatic testing allows financial and other savings to be made mainly because the human effort is not fragmented but is instead effectively concentrated and many tests can be made in a short time. The tests made are only valid if the program is well designed and it is possible to have reproducible tests even after a breakdown. Automatic testing is used by manufacturers of components such as integrated circuits because in some cases the cost of manual testing is prohibitive. Automatic testing allows the rapid selection of components for matching purposes and also makes it possible to test subassemblies of complex equipment. To this end DACs are used as programmable sources, ramp generators or generators of whatever other wave form is needed, or for controlling the gain of calibration loops while ADCs are used to convert analog measurements into digital form which are then compared with reference values stored in a memory.

1.2.2 Arithmetical and Trigonometrical Operations

The following operations can be carried out using converters.

Multiplication. If the reference voltage of a DAC is varied then the DAC will multiply; it can also be said that this arrangement is a variable gain amplifier. If the DAC is inserted into the feedback loop of an operational amplifier a division can be carried out.

Hybrid addition. When it is required to add two digital numbers they can for example be applied to two DACs supplying an output current and fed from the same reference voltage. The two output currents will be summed in a resistor and the resultant current will be proportional to the sum of the numbers.

Function generators. This category comprises all 'generators' producing a given function $y = f(x)$ which can be linear or not. When a linearly varying signal is fed to the input, the output will give a signal which varies according to the given function. In practice this is obtained by using circuit elements whose charac-teristics obey the required equation (for example a diode for an exponential function) or by using an arrangement of diodes and resistors the given function is approximated by a series of linear segments or by a combination of both methods. It is also possible to use digital memories (for example ROMs) to store a function in digital form and to use them with ADCs and DACs, one of the standard applications being the generation of simple trigonometric functions such as $\sin \theta$ or $\cos \theta$. Computer controlled programmable memories are used so as to be able to vary at will the simulated function.

1.2.3 Communications and Signal Analysis

This group includes converter applications dealing with time compression and time delay, the memory storage of transient phenomena, the synthesis of transfer functions, correlation and digital filtering, etc.

Analog delay lines. The signal that is to be delayed is first digitized using an ADC, each bit is then fed to a delay line consisting of a shift register and the sample proceeds at a given clock rate. n registers are required if the ADC supplies n bits. A DAC at the delay line output reconstructs a delayed analog signal, the delay being proportional to the number of shifts in a register and to the clock period.

Correlation. If a number of outlets are provided at regular intervals along the delay line just described, then by connecting a DAC to each outlet the function $f(t - \tau)$ will be available $f(t)$ being the input function and τ the delay between two outlets. Thus the operation $f(t)f(t-\tau)$ can be easily obtained for various multiples of τ, and this constitutes an elementary stage of correlation.

Transient analysis. When very short transient disturbances are to be studied, it is useful to record them in digital form by storing them in a shift register. In this way they can be regenerated at a slower speed and so observed more easily. This technique is used in certain types of oscilloscopes called transient recorders. It is possible to use a recirculating memory so as to obtain data repetitively from a single event though the number of samples observable is thereby reduced.

Averaging by addition. When a repetitive signal is accompanied by noise, it is possible to reduce the effect of the noise by adding up several periods of the signal.

The correlated parts add up directly whereas the noise only increases as the square root of the number of additions made. A delay line is used for this purpose and the incoming sample is added to the sum of corresponding samples which have preceded it, i.e. samples taken at the same instant of the period.

1.2.4 Visual Displays

On certain types of oscilloscopes the sensitivity and sweep rate setting are displayed on the screen and this requires the use of converters.

Another interesting application of visual display devices is in linking data acquisition systems and computers. The combination of computer and visual display unit (VDU) has great potential. In this case ADCs and DACs provide the link with the external world, DACs also being used for sweep generators, spot intensifiers, etc.

Television scanning. The horizontal scan is very fast whereas the vertical deflection time is slower so that the whole frame is written during the scanning period. The information is provided by modulating the current during the horizontal scan. This method is used for the display of alphanumeric texts in conjunction with a digital memory where the text to be displayed can be processed and stored. During the horizontal scan the memory controls the current modulation (usually ON/OFF: dark/bright spot). DACs can be used to obtain the spacing between lines, the line scan etc.

Matrix display. Another method of displaying a character is to locate the position it should occupy on the screen by means of its coordinates x and y, then taking that point as origin, the character is reconstructed by displaying a series of segments sequentially or continuously. This method of scanning is sometimes known as random scan. Usually a 35 point (5×7) matrix is required to display a character. DACs and ROMs are used to form and control these matrices.

Graphic display. It is desirable to be able to display information on a screen in numerical form or as images, and to be able to modify it, store it, separate part of it etc. These operations are very well handled by combining digital techniques and DACs the latter being used for the VDU deflection.

1.2.5 Industrial

This is the most important field of application, in terms of number of products and range of innovations. The first converters were developed to interface with computers. It would be wrong, though, to think that they are still restricted to this field of application alone. Just as was the case for the operational amplifier, $A \rightleftharpoons D$ converters have left the confines of the laboratory and entered the industrial field, finding a place each time it is advantageous for a given application to change from an analog to a digital form or vice versa. For example:

Digital communications. These have helped reduce significantly the influence of noise on the quality of the transmitted signals.

Automatic weighing systems. Converters are used for automatic tareing, and the weight of the container is converted into a digital signal by means of an ADC which is fed to a DAC, whose error signal output is used to set the ADC. The weighing operation can then proceed in the usual way the result being available in numerical form without the DAC contents being modified.

Telemetry systems. In these, large quantities of information are transmitted in digital form and converters are used to encode and decode that information. Telemetry systems are often used on aircraft.

These lists of examples are inevitably incomplete and are in no way restrictive. Their purpose is simply to try and convey the vast possibilities open to devices such as converters. With a little imagination anyone should be able to find many other applications within his or her own field of work.

1.3 OPERATIONS CARRIED OUT DURING CONVERSION

Information processing systems may be divided into analog systems and digital systems. In the first case the systems handle signals that vary in a continuous way whereas in the second case the systems handle discrete variables called digital numbers. Any digital treatment of an analog signal first requires an analog to digital conversion operation. If it is desired to recover the processed information in its initial analog form then the inverse conversion operation is needed.

It is necessary now to examine in some detail the operations required by the conversion processes.[2,3] Let $x(t)$ represent a given analog signal and let $x^*(t)$ be the series of discrete values of the given signal taken at regular time intervals of period T_e. The operation which allows the change from $x(t)$ to $x^*(t)$ is the *sampling operation*. Thus in an *analog to digital conversion* process a series of numbers $\{a_n\}$ is obtained which corresponds to the input signal $x(t)$, each number representing the amplitude of one of the samples forming the signal $x^*(t)$.

In order to carry out the conversion process satisfactorily it is often necessary to store the analog sample $x^*(t)$ and this is called the *hold operation*. The amplitudes of the stored samples are then converted into digital numbers. The amplitude of a stored sample can, *a priori*, have an infinite number of values, whereas the digital number can only have discrete values. It is therefore necessary to replace the exact amplitude of the sample by a whole number of quanta, or steps, such that the resultant value of the amplitude is as close as possible to that of the actual amplitude. This operation is called *quantization*. The number thus obtained is then expressed in coded form, for example in binary code. This is the *encoding*

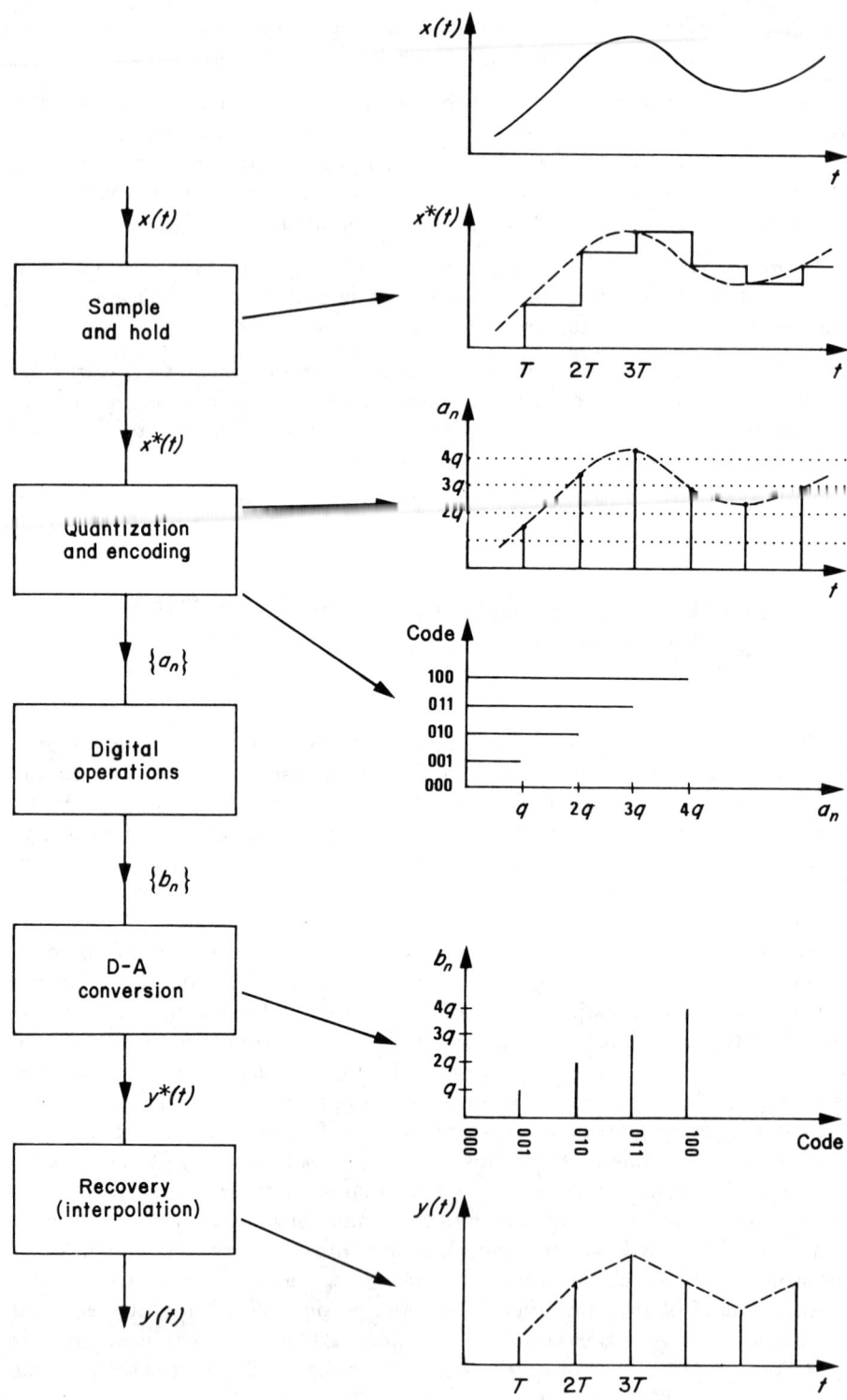

Figure 1

operation. The two operations, quantization and encoding, constitute the analog to digital conversion process.

Digital to analog conversion is rather simpler to carry out. The series of digital words obtained after some digital process is transformed into a series of discrete amplitude samples $y^*(t)$. To reproduce an analog signal output the values between samples must be obtained by interpolation. Often an interpolation of zero order is used followed by a low-pass analog filtering. This is the recovery operation. Figure 1 is a block diagram of these various operations and also shows the type of signal present at each stage of the process.

Each one of these fundamental operations will now be studied in detail in order to establish its main characteristics and be able to appreciate the role these characteristics could play in an A⇌D conversion problem.

1.4 THEORETICAL CONSIDERATIONS

1.4.1 Sampling

Sampling[4-7] is the first operation that may have to be carried out during a conversion process. To sample a function involves extracting periodically the value of that function over a certain time interval. It therefore amounts in effect to replacing the given continuous function by a discontinuous function consisting of 'cut outs' from the original function (Fig. 2). Provided certain conditions are satisfied the sampled function accurately describes the original function.

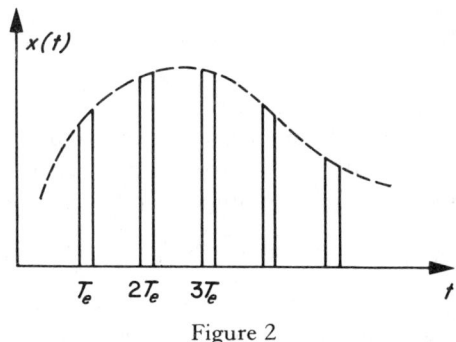

Figure 2

The important parameter in the sampling operation is the *sampling rate* which depends on the shape of the analog signal to be converted. In order to obtain the values that the sampling rate may take, it is necessary to carry out a spectral analysis of the sampled signal. First the concept of *ideal sampling* will be examined since the theory behind it is simple. In the case of ideal sampling it is assumed that the width of the sampling pulses is zero. Let $x(t)$ be the signal to be sampled and

$x^*(t)$ the sampled signal. The train of sampling pulses is represented by the function $u(t)$. A signal $x(t)$ is represented in the frequency domain by the function $X(f)$ such that $X(f)$ is the direct Fourier transform of $x(t)$, that is:

$$X(f) = \int_{-\infty}^{+\infty} x(t)\, e^{-2\pi jft}\, dt$$

The function $x^*(t)$ representing the sampled signal is given by:

$$x^*(t) = x(t)u(t)$$

For ideal pulses of unit amplitude, period T and zero width, $u(t)$ can be written as

$$u(t) = \sum_{n=-\infty}^{+\infty} \delta(t - nT)$$

Where $\delta(t)$ is the mathematical symbol used to represent an infinitely narrow pulse, the Dirac delta function, which is defined to have the following properties; $\delta(t) = 0$ for $t \neq 0$,

$$\int_{-\infty}^{+\infty} \delta(t)\, dt = 1$$

The function $x^*(t)$ is then given by

$$x^*(t) = x(t)u(t) = x(t) \sum_{-\infty}^{+\infty} \delta(t - nT)$$

In effect $x(t) \cdot \delta(t) = x(0)\delta(t)$ since $\delta(t) = 0$ for $t \neq 0$

Therefore the expression of the Fourier transform of the sampled signal which gives the spectrum of the signal is:

$$X^*(f) = \sum_{n=-\infty}^{+\infty} X(f - nF_e) \quad \text{where } F_e = \frac{1}{T}$$

The spectrum of the sampled signal can be traced out using this last equation and generally it will be markedly different from the spectrum of the original signal $x(t)$. To each frequency line of the spectrum of the function $x(t)$ corresponds a double infinity of lines of the same amplitude and of frequency

$$f' = f \pm nF_e \quad \text{(Fig. 3)}$$

It can be anticipated intuitively that the substitution of the sampled function for the original function may only be made with great care, for we must be able to identify the original function again. It can be seen from Fig. 3 that this sampling operation can be realized if a relationship exists between the sampling frequency F_e and the maximum frequency F_{max} of spectrum $X(f)$. If the signal was sampled at a frequency $F_e \geqslant 2F_{max}$ the series of spectra will not overlap and it is then possible to separate the central band by means of an *ideal low-pass filter*. This relationship between F_e and F_{max} is called the sampling theorem or *Shannon's theorem*. It may be stated thus; 'A signal $x(t)$ whose spectrum is limited to the

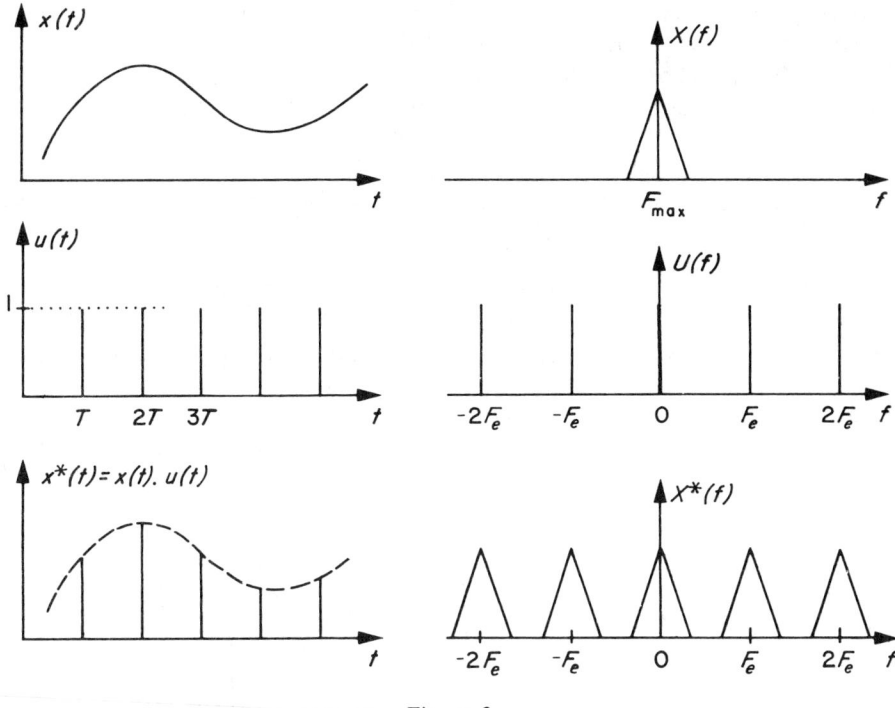

Figure 3

frequency $F_{\max}$ is completely determined by the entire series of its samples taken at regular intervals $T = (1/2F_{\max})$.'

If this condition is not satisfied, the successive spectra partially overlap and it is no longer possible to recover the original information. In practice two restrictions have also to be taken into account; the upper limit of the spectrum of signal $x(t)$ is not clearly defined and simply decreases above a given frequency, and an ideal filter does not exist so at the recovery stage a real filter must be used which will allow through some of the higher frequency bands. Therefore, in order to choose the sampling rate a knowledge of the shape of the spectrum of the analog signal is required and this reflects upon the speed of operation required of the converter.

In practice it is almost impossible to obtain ideal Dirac pulses and the pulses normally used always have a certain width τ. The way in which this parameter modifies the spectrum of the sampled signal must now be examined.[8] The first type of real sampling used is *analog sampling*; the pulses have a certain width τ and the signal $x^*(t)$ faithfully reproduces the signal $x(t)$ for the duration of these pulses. The Fourier transform of the sampled signal can then be written;

$$X^*(f) = \sum_{n=-\infty}^{+\infty} \frac{\sin \dfrac{n\pi\tau}{T}}{\dfrac{n\pi\tau}{T}} F(f - nF_e)$$

Each repeated spectrum is multiplied by a constant factor $(\sin x)/x$ which depends on n only. Therefore the central part of the spectrum of $X^*(f)$ is not distorted (Fig. 4). The overall width of the spectrum of such a signal is reduced compared to that obtained using ideal sampling and it can be assumed to be limited to $1/\tau$.

The second sampling method used is *sample and hold*. The pulse has a duration τ and a constant amplitude. In this case the Fourier transform of the sampled signal is given by:

$$X^*(f) = \frac{\tau}{T} \frac{\sin \pi \tau f}{\pi \tau f} \sum_{n=-\infty}^{+\infty} (f - nF_e)$$

The resultant spectrum is shown in Fig. 5 and in this case distortion does occur.

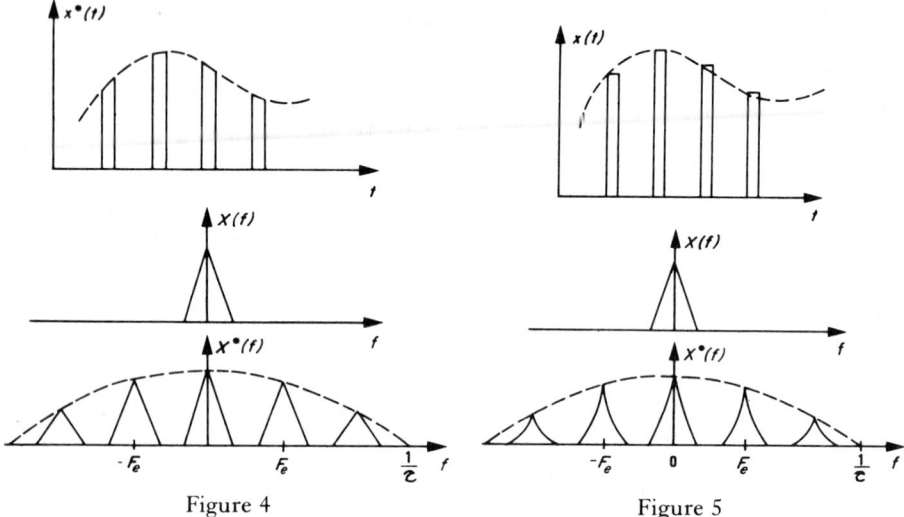

Figure 4   Figure 5

The error introduced when sampling a signal is an important problem since it affects the accuracy of analog to digital conversion processes. It is particularly significant when it is required to convert fast, discrete phenomena into digital form, but this problem is beyond the scope of this work and the interested reader is referred to more specialized works.[9,10]

1.4.2 Recovery

Following a digital to analog conversion operation there arises the problem of *recovery*,[11,12] for once the D–A operation is carried out the samples are re-established and a function $x^*(t)$ is obtained whose spectrum is $X^*(f)$. The original signal $x(t)$ which gave the train of samples must now be reconstructed or recovered.

Let us consider a signal $x(t)$ whose spectrum is limited to the frequency F_{max} and which is sampled at a rate $F_e = 2F_{max}$. It can be shown that $x(t)$ can be expressed in terms of its samples as;

$$x(t) = \sum_{n=-\infty}^{+\infty} x\left(\frac{n}{2F_{max}}\right) \frac{\sin(2\pi F_{max}t - n\pi)}{2\pi F_{max}t - n\pi}$$

If a sample $x(n/2F_{max})$ is selected and is fed through an ideal low-pass filter with cut-off frequency F_{max} the response of the filter to such an input pulse will be;

$$y_n(t) = 2F_{max} x\left(\frac{n}{2F_{max}}\right) \frac{\sin(2\pi F_{max}t - n\pi)}{2\pi F_{max}t - n\pi}$$

The response of the filter to all the samples gives an output signal;

$$y(t) = \sum y_n(t) = 2F_{max}x(t)$$

Ignoring the constant coefficient, this expression is identical to the original signal $x(t)$. It is therefore possible to recover the original signal $x(t)$ from samples taken at time intervals $1/2F_{max}$ by passing them through an ideal low-pass filter having a cut-off frequency of F_{max}. However in practice such a filter does not exist.

Nevertheless in order to recover the signal a classical *low-pass filter* may be used, for example a Butterworth or Tchebycheff type of filter which have the advantages of having simple transfer functions and being easily realizable. The fidelity of the recovery will depend on the ratio F_c/F_e, F_c being the cut-off frequency of the filter, and on the sharpness or slope of that cut-off. The following expressions are for evaluating the error in recovery when a Butterworth filter is used. Consider a train of samples $x^*(t)$ which is passed through a filter having a transfer function $H(f)$ (the transfer function of a system is the relation between the input and output signals of that system). The Fourier transform $Y(f)$ of the output signal is given by

$$Y(f) = H(f) \cdot X^*(f)$$

The quality of the recovery may be evaluated by calculating the mean square value of the error

$$\varepsilon(t) = y(t) - x(t)$$

which can be expressed in terms of the Fourier transforms (Parseval theorem):

$$\overline{\varepsilon^2}(t) = \int_{-\infty}^{+\infty} |y(t) - x(t)|^2 \, dt = \int_{-\infty}^{+\infty} |Y(f) - X(f)|^2 \, df$$

For a Butterworth filter this error is then given by:

$$\overline{\varepsilon^2} = 2^{2n}\left[\frac{n}{\pi(2n-1)}\left(\frac{F_c}{F_e}\right)^{2n-1}\sin\left(\frac{\pi}{2n}\right)\right]$$

Another solution makes use of an *interpolator*. An interpolator is a device which allows the original curve to be reconstructed from the samples using linear segments. This operation always involves an error because of the approximate nature of the interpolation function used. The simplest type of interpolation consists of holding the signal constant between two consecutive samples. This method is analogous to the hold device used in the 'sample and hold' method of sampling. This operation is called *zero order interpolation* or *'stepped' interpolation*. The transfer function of such an interpolator is given by:

$$H(f) = T \frac{\sin \pi Tf}{\pi Tf} e^{-j\pi Tf}$$

Its impulse response is a unit pulse of width T. The spectrum of the sampled signal is distorted since it is multiplied by the above function. The amplitude is multiplied by a function of $(\sin x)/x$ and there is also a linear phase shift of $-\pi fT$. This distortion is significant even if the sampling theorem has been amply satisfied.

A more interesting solution makes use of linear interpolation; that is first order interpolation. The signal between time intervals nT and $(n+1)T$ is a function of samples

$$x(nT) \quad \text{and} \quad x((n-1)T)$$

and can be expressed on

$$x(nT + \theta) - x(nT) + \frac{k\theta}{T}(x(nT) - x[(n-1)T])$$

with

$$0 < \theta < T$$

k is positive and less than 1. When k is suitably chosen (of the order of 0.3 to 0.4) then a frequency response curve for the interpolator can be obtained which is almost flat within the band $(0, 1/2T)$ which is the useful band. However, as in the case of the zero order interpolation a certain phase shift will occur.

1.4.3   Quantization[13–15]

Quantization[13–15] is an operation that is met in systems used for transforming an analog quantity into digital information in the form of a message with a well defined length. Thus for a base b, n bits will allow $N = b^n$ distinct values to be encoded. It is possible to encode, that is to say recognize, only N well defined values of the analog quantity.

The quantization operation consists in replacing the exact value of the signal by one value taken from the N quantized values. Thus information consisting of a finite number of values is substituted for the possible infinite number of values of the analog quantity. Therefore it is essentially a nonlinear operation. In order to quantize a signal the amplitude axis is divided into levels which are assigned numbers. It is then only necessary to indicate the level number in which the tip of the vector representing the signal lies in order to define fully the amplitude of the signal. Therefore in order to carry out a quantization, the interval between levels must be defined. This interval is called the *elementary level of quantization* or *quantum q.* Any signal whose amplitude lie in the interval $(nq, (n+1)q)$ will be described by the quantity nq.

In the simplest case the quanta are all equal. The way the output signal varies as the input signal is varied is shown in Fig. 6, together with the quantization error. The input–output curve is usually called the *quantization characteristic.* The accuracy of the quantized signal will depend upon the smallness of the chosen quantum. Figure 6 shows that any quantization introduces an error because the exact value of the signal is replaced by an approximate one (in other words an infinite number of values is replaced by a finite number of values).

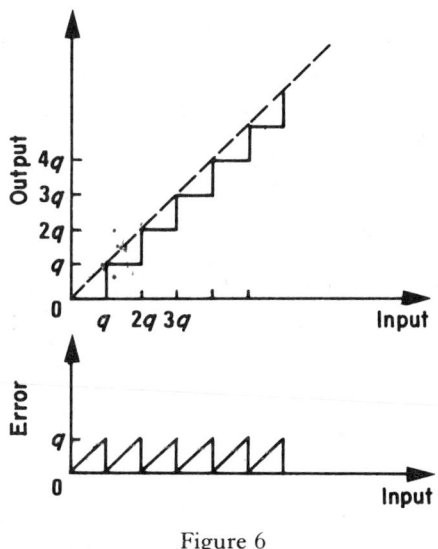

Figure 6

The absolute error is always smaller than the quantum q i.e. $|\varepsilon| \leqslant q$. When the digital information is given as a word of n bits in a system of base b, and if the maximum value of the amplitude of the signal to be converted is E the quantum level is E/b^n and the error satisfies the relation; $0 < |\varepsilon| \leqslant E/b^n$ and the maximum relative error is $1/b^n$.

In practice by ensuring that the value of the output signal changes when the input signal exceeds one of the $(2n+1)q/2$ values the error is reduced to $\pm q/2$. The quantization error is also called the *quantization noise* because it often has the same effect as random noise.

When the input signal varies with time the curve representing the quantized signal has the shape of a *'stepped'* or *'staircase' curve* (Fig. 7) and the corresponding error $\varepsilon(t)$ depends upon the elementary level of quantization and upon the input signal $x(t)$. When the elementary level q is sufficiently small compared to the amplitude range of the signal $x(t)$ the error may then be considered as noise whose frequency, related to the maximum slope of the curve $x(t)$, depends upon the highest frequency of the spectrum of the input function. In other words as a first approximation the statistical distribution of the quantization error can be taken as

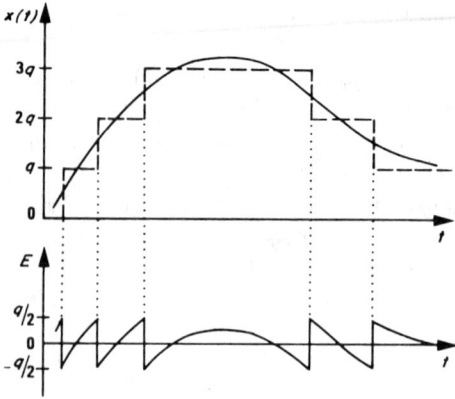

Figure 7

uniform. The error $\varepsilon(t)$ can be regarded as a series of linear segments of variable slope between the limits $\pm q/2$ (Fig. 7) except where the signal passes through a maximum or a minimum.

If α is the variable slope of these segments during a time interval

$$-\frac{q}{2\alpha} < T < \frac{q}{2\alpha}$$

the error may then be expressed as

$$\varepsilon = \alpha t$$

The mean square error (normalized energy of the quantization noise) is given by

$$\overline{\varepsilon^2} = \frac{\alpha}{q} \int_{-q/2\alpha}^{q/2\alpha} (\alpha t)^2 \, dt = \frac{q^2}{12}$$

In order to proceed with quantization the coding system which will enable the quantized signal to be represented by a word must be defined and the base of the code must be selected. For reasons of convenience the base customarily chosen is the base of 2, the code being the *binary code*, since many components exist which can have two stable states.

In this case the quantization error becomes $1/2^n$ and the signal-to-noise ratio (S/N) for the quantization can be calculated. The maximum number of distinct states is $N = 2^n$. If E is the maximum value of the input signal the ratio (S/N) expressed in the binary system has the value:

$$\left(\frac{S}{N}\right) = \frac{E^2}{\overline{\varepsilon^2}} = 12\frac{E^2}{q^2} = 12N^2$$

$$\left(\frac{S}{N}\right) = 12 \cdot 2^{2n}$$

or

$$\left(\frac{S}{N}\right)_{dB} = 10.8 + 6n$$

Therefore the signal to noise ratio increases by 6 dB each time an extra bit is added, that is each time the quantum value is divided by two. One of the disadvantages of linear quantization is that when the amplitude of the signal varies the value of the signal-to-noise ratio also varies, since the quantum remains constant. To overcome this difficulty a *non linear* type of quantization is used and this frequently takes the form of a logarithmic quantization. In this case the quanta are no longer equal but the value of each quantum is a function of its rank, that is:

$$q_i = \sigma \log i \quad \text{(for a quantum of rank } i\text{)}$$

When such a quantization is used it is possible to obtain a constant signal-to-noise ratio over a large range of amplitudes (say 30 dB) of the input signal and this is used in telecommunications.

1.5   CODES

The last stage of an analog to digital conversion process is that of encoding, which establishes a one to one correspondence between the number of quanta expressing the amplitude of a sample and its translation into a specific code. The codes commonly used are binary codes, though many forms of binary codes are in existence. Selecting the code to be used is an important step for it affects the performance of the converter. The two states of the binary digit or bit are represented by 0 and 1; a 'one' indicates that the bit is to make a contribution towards the numerical count and a 'zero' that it is not. Normally a number is made up of n digits, each one contributing b times more towards the numerical count than the previous digit on the right, b being the base system used. For example in the decimal system there are units, tens (worth ten units), hundreds (worth ten tens) etc. Each digit state has an associated voltage value (since circuits for logic operations use voltages and not numbers).

Two types of logic may be defined depending upon the values given to the associated voltages; positive logic, for which the state 0 of the bit corresponds to zero voltage and state 1 to +5 V, and negative logic for which the reverse conventions are used.

These are theoretical values. Nowadays logic circuits predominantly use transistor transistor logic (TTL) circuits and state 1 corresponds to a minimum output level of 2.4 V and state 0 to a maximum level of 0.8 V. The binary codes used may be divided into two groups according to whether the analog signal to be encoded is always of the same sign or is perhaps alternately positive and negative, thus defining unipolar and bipolar codes.[3]

1.5.1 Unipolar Codes

The best known is the *natural binary code* also called pure binary. In such a code a number N is expressed as a weighted sum:

$$N = \sum_{i=0}^{n-1} a_i 2^i = a_{(n-1)} 2^{(n-1)} + a_{(n-2)} 2^{(n-2)} + \ldots + a_0 2^0$$

The coefficient a_i is 0 or 1 depending on whether the corresponding bit is zero or not. Fractional numbers may also be expressed in this way:

$$N' = a_1 2^{-1} + a_2 2^{-2} + \ldots + a_n 2^{-n}$$

Expressing a number in binary is done therefore by writing the number in decreasing powers of 2. Thus the decimal number 45 is written as

$$45 = 1 \times 2^5 + 0 \times 2^4 + 1 \times 2^3 + 1 \times 2^2 + 0 \times 2^1 + 1 \times 2^0$$

that is,

$$N = 1\ 0\ 1\ 1\ 0\ 1$$

and for the fractional number 45/64

$$N' = 1\ 0\ 1\ 1\ 0\ 1$$

A voltage V may be assigned to this number N' such that

$$V = V_{ref} \left(\frac{a_1}{2} + \frac{a_2}{4} + \ldots + \frac{a_n}{2^n} \right)$$

V_{ref} being a constant voltage used as a scale factor.

The bit of greatest weight a_1 is called the most significant bit (MSB) and the bit of least weight, a_n is the least significant bit (LSB). The quantum (or elementary step) previously defined as $q = E/2^n$ is obtained when all the bits except a_n are in the zero state, a_n being in the state 1. The bit of least weight is therefore the digital equivalent of the quantum.

A second unipolar code is *binary coded decimal* (BCD) in which each decimal number is represented by a binary word of four bits. The corresponding voltage can be written as

$$V = \frac{V_{ref}}{10} (8a_1 + 4a_2 + 2a_3 + a_4) + \frac{V_{ref}}{100} (8b_1 + 4b_2 + 2b_3 + b_4) + \ldots$$

The groups $(a_1 a_2 a_3 a_4)$, $(b_1, b_2, b_3, b_4) \ldots$ corresponding to the binary translation of the decimal figures. The 1–2–4–8 or 1–2–4–2 codes may be used to obtain this translation. Analog digital converters using such a BCD code are to be found mainly in digital voltmeters and display equipment because this code makes it possible to use a simple decoder.

A third unipolar code is the *Gray code* also called the *reflected binary code*. The attraction of this code is due to the fact that the passage from one number to the next consecutive number requires only one bit to be changed, so that intermediate erroneous changes can be avoided. This problem will be discussed later when digital to analog converters are considered. The code is often used for angular encoding systems. The translation from natural to reflected binary is made as follows; the MSB is preserved then reading the natural binary from left to right each change is a '1' in reflected binary and lack of change is '0'. For example, the natural binary 1 0 1 1 is translated into Gray code as 1 1 1 0.

The translation from natural to reflected binary can be carried out by the circuit given in Fig. 8 which uses $n - 1$ exclusive OR gates. (The exclusive OR (XOR) logic circuit has the following characteristics; the output signal has the value one if one and only one of its inputs is in the state one, otherwise it is zero.) The first 15 consecutive numbers expressed in the above three codes are given in Table 1.

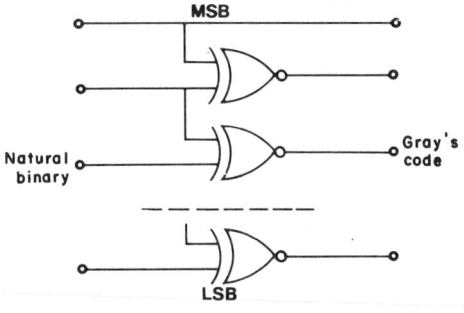

Figure 8

Table 1. The first fifteen numbers expressed in three different codes

N	Natural binary	BCD 8–4–2–1	Gray's code
15	1111	0001 :0101	1000
14	1110	0100	1001
13	1101	0011	1011
12	1100	0010	1010
11	1011	0001	1110
10	1010	0001:0000	1111
9	1001	1001	1101
8	1000	1000	1100
7	0111	0111	0100
6	0110	0110	0101
5	0101	0101	0111
4	0100	0100	0110
3	0011	0011	0010
2	0010	0010	0011
1	0001	0001	0001
0	0000	0000	0000

1.5.2 Bipolar Codes

The signals to be converted most of the time may have positive or negative amplitudes and it must be possible to distinguish between two signals of equal amplitude but of opposite sign. Under these circumstances a bipolar code is used. The first possible bipolar code is the *sign and magnitude code* also called the *symmetric binary code*. It appears to be the most straightforward method of encoding analog signals taking into account their signs, yet it is little used because the numbers obtained do not lend themselves well to arithmetical operations. To obtain such a code the amplitude is encoded separately in natural binary and the sign is indicated by an additional bit which is 1 for positive signals and 0 for negative signals. This code has advantages over other bipolar codes when used for signals varying closely about zero for in all other codes crossing through zero can cause all the bits to change which may result in errors or in severe transient states. A number represented in natural binary by n bits will be represented by $n + 1$ bits in a bipolar code, but the quantum remains the same since the full scale has now doubled.

The *shifted* or *offset binary code* is a natural binary code that has been offset so that the number zero is produced for the most negative voltage $-V_{ref}$ and the number corresponding to 2^{n-1} (mid-scale) is obtained for zero volts. A voltage V may be expressed in offset binary as:

$$V = 2\frac{V_{ref}}{2}(a_1 2^{n-1} + a_2 2^{n-2} + \ldots + a_0) - V_{ref}$$

Mathematically a number N can be expressed in offset binary starting with its value in pure binary using the relation

$$N_{ob} = \pm N_b + 2^n$$

where n is the number of bits in the binary word N_b. For example for $n = 3$ the number $+5$ is written in offset binary as

$$N_{ob}(+5) = 101 + 1000 = 1101$$

and -5

$$N_{ob}(-5) = -101 + 1000 = 0011$$

In this code the sign bit (i.e. the MSB) has value 1 for positive numbers and 0 for negative numbers.

Two's complement code is very much used when representing bipolar signals for it lends itself very well to arithmetical operations, subtraction being replaced by the addition operation. Two's complement code may be obtained from the offset binary code by inverting the most weighted bit (the sign bit). Thus:

$$V = 2\frac{V_{ref}}{2^n}(\bar{a}_1 2^{n-1} + a_2 2^{n-2} + \ldots + a_0) - V_{ref}$$

In two's complement code a positive number is expressed by its binary equivalent and the representation of a negative number is obtained by taking the complement of each bit of the corresponding positive number and then adding 1 to the least significant bit. For example the number -5 is formed as follows:

$$0101 \rightarrow 1010 + 0001 = 1011$$

Mathematically this is written as

$$N_{\text{comp2}} = 2^{n+1} - N_b$$

Lastly we have *one's complement code* used, for example, in counters. The representation of a negative number is obtained by taking the complement of all the bits of the corresponding positive number. This may be written:

$$N_{\text{comp1}} = 2^{n+1} N_b - 1$$

For the number -5 this gives

$$101 \rightarrow 1000 - 00101 - 00001 = 1010$$

Table 2. The numbers from -8 to $+7$ in four different codes

N	Sign magnitude	Offset binary	2s complement	1s complement
7	1111	1111	0111	0111
6	1110	1110	0110	0110
5	1101	1101	0101	0101
4	1100	1100	0100	0100
3	1011	1011	0011	0011
2	1010	1010	0010	0010
1	1001	1001	0001	0001
0	1000	1000	0000	0000
-1	0001	0111	1111	1110
-2	0010	0110	1110	1101
-3	0011	0101	1101	1100
-4	0100	0100	1100	1011
-5	0101	0011	1011	1010
-6	0110	0010	1010	1001
-7	0111	0001	1001	1000
-8	—	0000	1000	—

Table 2 lists the encoded values of numbers between -8 and $+7$ in all the four codes (with n limited to 4 the number $+8$ cannot be encoded). Referring to the expressions for the reference voltage V corresponding to a number N relating to the various codes, it can be seen that the relationships existing between the maximum value of $V(V_{\text{max}})$, the reference voltage V_{ref}, the quantum and the possible range of variation of V ($V_{max} - V_{min}$) depend upon the code used, thus:

$$q = \frac{V_{ref}}{2^n}$$

$$V_{max} = \frac{V_{ref}}{2^n} (2^n - 1)$$

$$V_{max} - V_{min} = \frac{V_{ref}}{2^n} (2^n - 1)$$

For a unipolar code:

$$q = 2 \frac{V_{ref}}{2^n}$$

$$V_{max} = 2 \frac{V_{ref}}{2^n} (2^n - 1) - V_{ref}$$

$$= \frac{V_{ref}}{2^n} (2^n - 2)$$

and for a bipolar code:

$$V_{max} - V_{min} = 2 \frac{V_{ref}}{2^n} (2^n - 1)$$

To end this description of the various types of codes it is of interest to summarize how a code may be chosen for a given application. The sign and magnitude code will only be used if high accuracy is required in the neighbourhood of zero. The offset binary code is often used in circuits containing summing amplifiers, as such circuits can easily cope with offset voltages or polarization currents. Two's complement code is often used in digital AC converters or in sequential converters, but one's complement code is only used in counters.

1.6 SAMPLE AND HOLD CIRCUITS

The conversion of a sample takes a certain length of time and it is important that the signal does not vary during the conversion operation otherwise its significance will be completely lost. It is therefore necessary for the amplitude of the sample to remain constant during the conversion time. This is achieved by a hold circuit inserted after the sample circuit.[16] It consists as a rule of a capacitor which is charged according to the value of the sample. A simple example will demonstrate the usefulness of this hold circuit.

Let $x(t) = E \cos \omega t$ be the sinusoidal signal to be converted, E being the maximum value to be encoded by the converter. The maximum rate of change of this signal is given by:

$$\frac{dx}{dt}\bigg|_{max} = E\omega = 2\pi fE$$

A storage operation prior to conversion is only essential if, during the conversion time t_c, the change of amplitude of the signal exceeds one quantum q. The maximum frequency allowable without using a hold circuit is given by:

$$f_{max} = \frac{q}{t_c} \cdot \frac{1}{2\pi E}$$

If an n bit encoding is used

$$2E = 2^n q$$

and

$$f_{max} = \frac{1}{t_c \pi 2^n}$$

If $n = 8$ and $t_c = 1$ ms a hold circuit is required for a signal frequency exceeding 1.2 Hz.

Sampling the analog signal and 'holding' the sampled values are operations that are required for practically any conversion process. Sample and hold circuits are not part of the converter proper and will only be described briefly. The operation of an ideal sampling device will be reviewed first, then the sources of error which are met in practice will be examined and finally some examples of simple circuits will be given.

1.6.1   The Operation of an Ideal Sampling Device

A sample and hold module has an analog input for the signal to be sampled, an analog sampled output and a digital control input. It has two operating modes, *sampling* and *holding* and changes from one to the other are effected by the control signal (Fig. 9).

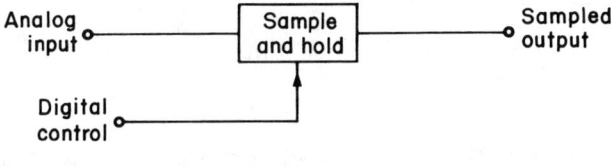

Figure 9

During the sampling time, which usually corresponds to a logic state 1 applied to the control input, the output voltage follows faithfully the changes of the input signal. When the control signal changes and takes the logic state 0, the output signal remains constant and equal to the last value of the input signal received.

When the control signal returns to logic state 1, the value of the output voltage changes instantaneously and becomes equal to the value of the input signal at that instant of time. This operating sequence is summarized in Fig. 10.

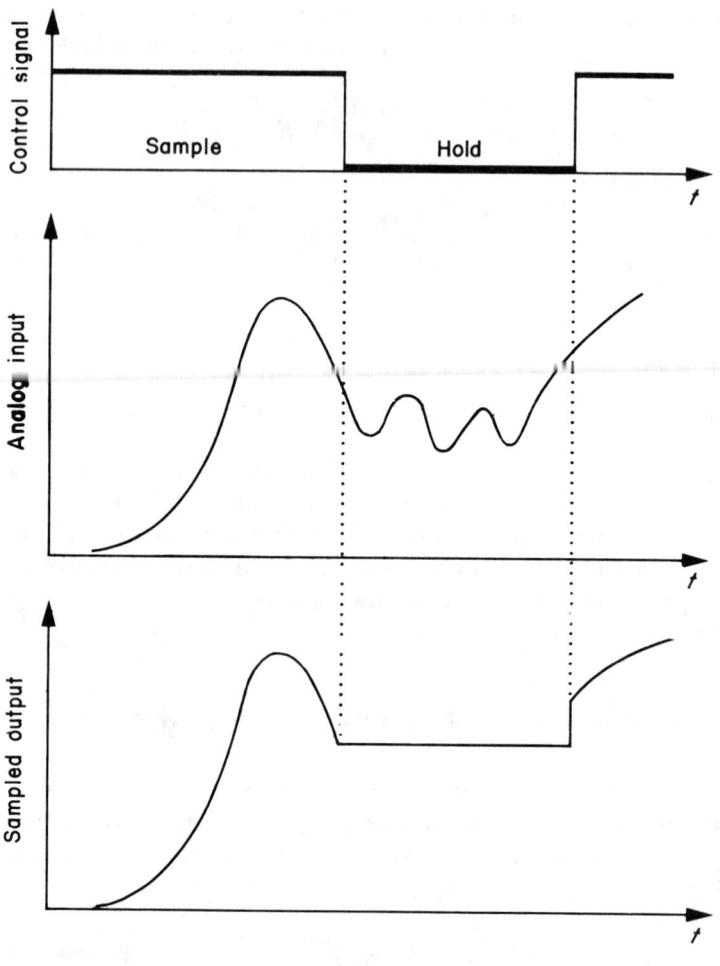

Figure 10

In order to carry out these various operations the basic circuit of a sample and hold device includes an analog switch operated by the command signal, a capacitor which holds the output signal during the logic state 0, and an operational amplifier to isolate the capacitor holding the charge. Such a sampling circuit is ideal only if its switching times are zero and there is no drift or leakage. This is not the case in practice and it becomes necessary to define the various likely errors related to the various operating modes of the system.

1.6.2 Errors in Sample and Hold Circuits

The errors met in practice can be defined in relation to four periods of time corresponding to the various operating modes of the elements of the sample and hold module. These are:

sample,
the transition from sample to hold,
hold, and
the transition from hold to sample.

1.6.2.1 Sample

The errors met during this period are essentially the offset error and the gain error (Fig. 11).

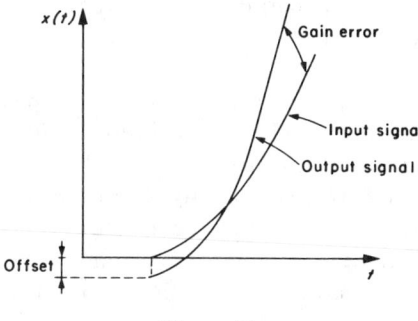

Figure 11

Offset error. This occurs when the input signal is zero and there is nevertheless an output signal. This is often due to the offset voltages of the amplifiers and can be compensated for by means of an external potentiometer. This error is also called zero setting error.

Gain error (also called error in scale factor). This occurs when the output signal is not equal to the input signal but only proportional to it. This is due to gain errors of the amplifier. It is usually compensated for by adjusting the overall gain of the system.

There is also the phenomenon of *dielectric absorption* in the storage capacitor. When an initially charged capacitor is discharged it will gradually recover part of its previous charge. It is therefore preferable to use polystyrene or polycarbonate type capacitors which have relatively low dielectric absorption coefficients.

1.6.2.2 The transition from sample to hold

When the control signal changes from 1 to 0, the output signal does not 'freeze' instantaneously but continues to change, thus introducing errors (Fig. 12).

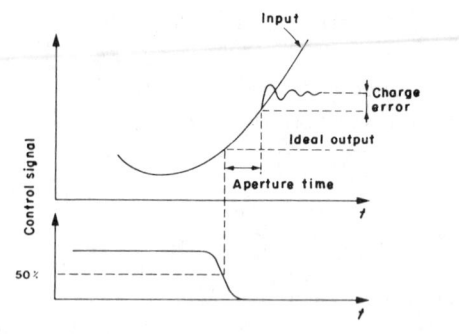

Figure 12

Aperture time (sample to store time). This is the time elapsed between the instant the command signal is given and the instant the switch is effectively opened stopping the capacitor charging. The charge stored in the capacitor differs from the desired value which is that of the signal at the instant of time of the command. This charge remains constant provided there is no error during the hold period. The aperture time is measured taking as the origin of the time scale the instant of time when the command signal reaches 50% of its value. This time greatly depends upon the performance of the switch. If great accuracy is required for the sampling time then it is necessary to advance the operation of the switch in order to reduce this delay. In practice the aperture time is not constant and is subject to some uncertainty. It is very important to know the degree of this uncertainty as it limits the accuracy to which the amplitude of the sample may be known.

Charge error. At the start of the hold period when the switch is turned off another error occurs due to the transfer of the capacitive charge of the switch drive circuit (for example the gate drain capacitance if a field effect transistor (FET) is used), into the storage charge of the capacitor. As a result the output signal has a 'step' which may be reduced, though not completely eliminated, by using a low capacity input FET and by increasing the storage capacity. Moreover transient states (due to overshoot) may occur which result in an extra period of acquisition time.

1.6.2.3   Hold

Dynamic non-linearity is a term often used to describe the errors that occur during the hold period. They include essentially the variation of the charge of the capacitor (drift) and the voltage excursion due to circuit coupling (Fig. 13).

Charge variation. The output voltage does not remain constant during the hold period but varies approximately linearly (at a constant rate). This corresponds to the linear variation of the charge resulting from a constant current flow:

$$\frac{dv}{dt} = \frac{I}{C}$$

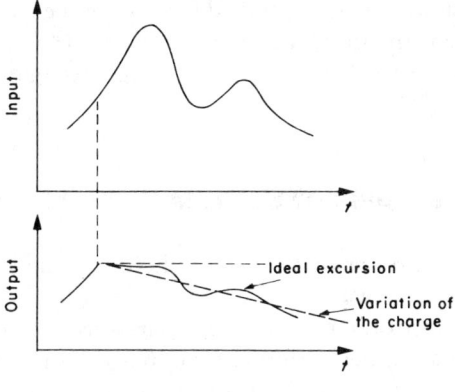

Figure 13

This current comprises the leakage of currents of the switch and of the capacitor, the offset current of the amplifier etc. The slope obtained may be positive or negative.

Excursion (feed forward). This is the fraction of the input signal which is found at the output. This error is due entirely to the capacitive coupling between the input and output of the switch.

The *dielectric absorption effect* is again encountered here.

1.6.2.4 The transition from hold to sample

In this case the important variable is the *acquisition time*. This is the time taken for the output signal to become equal to the input signal within a specified accuracy. When the control signal changes from 0 to 1, the output signal cannot instantaneously take the value the input signal has at that instant of time, as time is needed for the capacitors to become charged (Fig. 14). In order to reduce this charging time it is necessary to use a switch having a very low series impedance and to use a source of very low internal impedance while the buffer amplifier must have a very fast response.

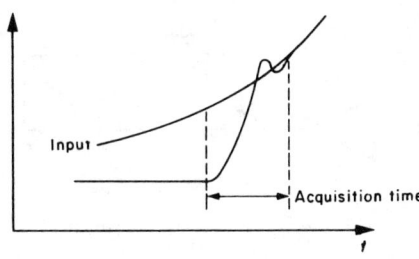

Figure 14

The acquisition time is usually specified for a full-scale change of the input and therefore to measure it the input signal must change from zero to its maximum value. As in the case for the transition from sample to hold oscillatory transient states may occur.

1.6.3   Some Examples of Sample Circuits

Some typical circuit diagrams of sample and hold circuits are shown in Fig. 15. The simplest of them consists of a capacitor which is charged using a switch (Fig. 15(a)). In Fig. 15(b) the capacitor is isolated from the circuit that follows by means of a buffer amplifier. A buffer amplifier has the following characteristics; its input impedance is very high (it can very often be taken as infinite), and its voltage gain is unity, that is the output voltage follows the changes of input voltage.

In the above two circuits the source is loaded during the sample period by a large capacitance which may result in oscillations. Therefore in Fig. 15(c) a buffer amplifier has been added at the input to isolate the source from the analog store.

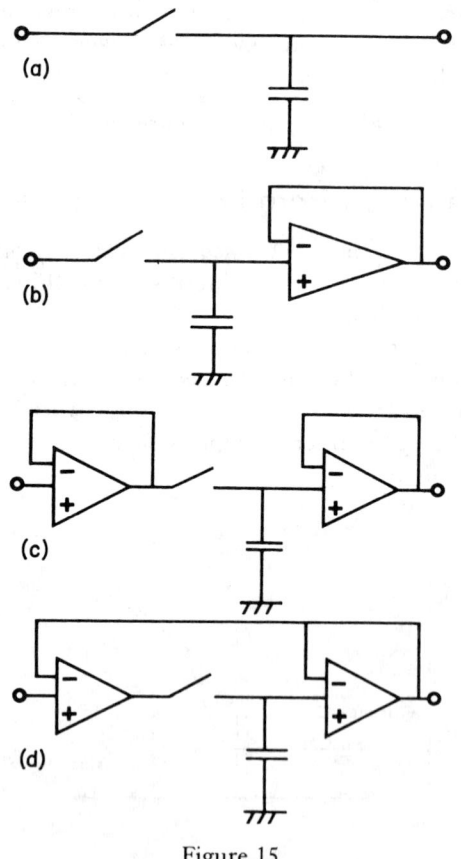

Figure 15

With this arrangement it is possible to achieve acquisition times of the order of only ten nanoseconds and the uncertainty over the switching time can be lowered to one ns. When the acquisition time is not critical a circuit such as that of Fig. 15(d) is of interest because a greater system accuracy is achieved by means of the feedback loop. In this case the offset error and the common mode operation of the buffer amplifier are automatically compensated by the adjustment of the charge stored on the capacitor. The switching function is increasingly being carried out by MOS type FETs which usually require interface circuits in order to drive them.

As an example, Fig. 16 shows a schematic diagram of a sample and hold circuit in common use. This circuit consists of two operational amplifiers with low offset currents, a hold capacitor, two FET switches and their drive circuit. This arrangement has a gain of unity and operates as a buffer arrangement.

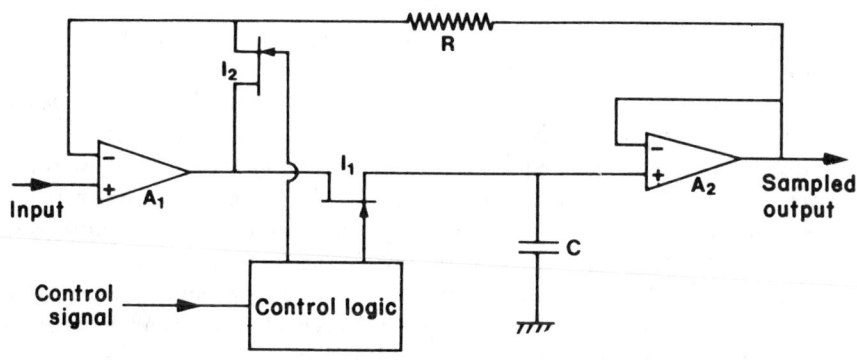

Figure 16

In the *sample mode* the drive circuit opens switch I_2 and closes switch I_1. The output voltage is then equal to the input voltage. The output current of the operational amplifier A_1 charges capacitor C until its voltage is equal to the input voltage. In the *hold mode* I_1 is opened and I_2 closed. The voltage of the charged capacitor shows practically no decrease because it 'sees' on the one side the input impedance of the FET operational amplifier connected as a buffer and on the other side an open FET switch. When switch I_2 is closed it prevents amplifier A_1 saturating. Since A_1 operates as a buffer the voltages across the drain and the source of the FET I_2 are equal. Resistor R limits the current flowing between the two amplifiers during the hold mode. The voltage drop across the terminals of this resistor is equal to the difference between input and output voltages.

2 DIGITAL TO ANALOG CONVERSION

2.1 DEFINITION

A *digital to analog converter* or DAC is a device which receives digital information in the form of an n bit word and transforms it into an analog signal.[2,17-20] It is therefore a hybrid system. A DAC causes one of the 2^n possible *binary combinations* being input (corresponding to an n bit input signal) to give rise to one of the 2^n *discrete voltages* obtained from a reference voltage V_{ref}. The law for the correspondence may be of any type but as a rule the natural binary is used together with a linear variation. There are many types of DACs; some give a voltage output and others a current output, some have an internal reference source whereas for others it is necessary to provide an external reference source, some give a unipolar output voltage while others accept bipolar codes. In certain cases the input signal to be accepted may be serial digital words on a single line whereas in other cases parallel digital words on separate lines are accepted. After defining the characteristic parameters of a DAC and the errors which may occur, the best known and most used present day types of DAC will be described.

The whole number N that is to be converted (or decoded) is expressed in terms of whole powers of 2 as follows:

$$N = d_1 2^{n-1} + d_2 2^{n-2} + \ldots + d_n 2^0$$

and its value will be between 0 and $2^n - 1$ depending upon the values of the coefficients d_i. To talk of the series of coefficients d_i is to talk about the binary number N, it being understood that it is expressed in base 2. Fractional numbers may also be decoded by expressing the numbers in terms of weighted coefficients corresponding to the negative powers of 2; that is a number less than one will be written as:

$$N' = d_1 2^{-1} + d_2 2^{-2} + \ldots + d_n 2^{-n}$$

or

$$N' = 0, d_1, d_2, \ldots d_n$$

which can also be written

$$N' = \frac{N}{2^n}$$

The value of N' is between 0 and 1.

The *transfer function* for a DAC can now be defined: if N' is the coded number which is fed to the DAC and V its output voltage then:

$$V = N' V_{ref}$$

or

$$V = \frac{N V_{ref}}{2^n}$$

that is

$$V = d_1 \frac{V_{ref}}{2} + d_2 \frac{V_{ref}}{4} + \ldots + d_n \frac{V_{ref}}{2^n}$$

V_{ref} is a voltage applied to the converter which serves as a reference for expressing V. V_{ref} determines the scale of the output voltage. The quantity $V_{ref}/2^n$ is the smallest detectable quantity and is called the *quantum*. From the above expression it can be seen that a DAC carries out the multiplication of the number N by the quantum. Hence an equivalent schematic diagram of a DAC can be deduced (Fig. 17). It has n logic inputs for the n bits of the word to be converted and an analog output (in this case a voltage). In addition a reference voltage V_{ref} must be applied to the DAC.

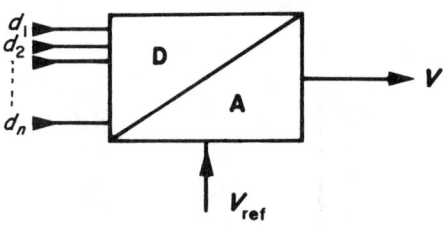

Figure 17

On the schematic diagram of Fig. 17 the converter's separate input lines are assumed to be in parallel, that is, all the bits are applied to the converter simultaneously. This assumption will be retained throughout the study of the various types of DACs, in order to simplify the treatment. In fact the binary word may be applied as a serial digital word and in order to change it to a parallel digital word a buffer memory (such as a shift register) must be used to store the information to be converted. At the start of the conversion process this information, now a parallel digital word, is applied to the input of the DAC in order to be converted. Shift registers will be discussed later on in the book.

If the voltage V_{ref} is constant then the DAC is known as a classic type of DAC. On the other hand if that voltage varies then the DAC is referred to as a *multiplying converter* or more simply a *D–A multiplier*.

2.2 THE DIFFERENT TYPES OF DACs

Different classifications of DACs can be made irrespective of the nature of the output signals (current or voltage) or of the way the digital information is applied to the input of the DAC. DACs can first be grouped into direct or indirect converters. In a *direct DAC* the binary word is converted into the output signal without an intermediary; in an *indirect DAC* conversion is obtained through the intermediary of an analog quantity such as time or a probability density.

A second possible classification is based on the different expressions of the transfer function given above. The voltage V can be considered as the *sum of elementary voltages* when it is written:

$$V = N\frac{V_{ref}}{2^n}$$

Equally, the *sum of weighted voltages* may also be considered by writing:

$$V = d_1\frac{V_{ref}}{2} + d_2\frac{V_{ref}}{4} + \ldots + d_n\frac{V_{ref}}{2^n}$$

(a) Direct sum

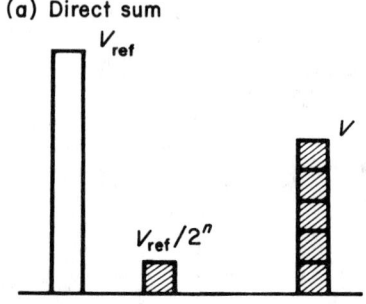

(b) Weighted sum

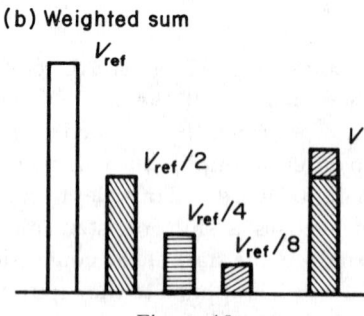

Figure 18

Figure 18 illustrates these two definitions when $n = 3$ and $N = 5$ and the expression for N' in this case is:

$$N' = 1 \times \tfrac{1}{2} + 0 \times \tfrac{1}{4} + 1 \times \tfrac{1}{8}$$

Normally the weighted sum is used for direct type conversions and the sum of elementary voltages is used for indirect type conversion.

2.3 CHARACTERISTIC PARAMETERS OF A DAC

It is important to specify carefully the different characteristics of a DAC for it is the values of these characteristics that will be quoted by the manufacturer. Unless otherwise stated these characteristics will be specified for unipolar codes.

The first characteristic is the ideal *transfer function* (error free) of a DAC. It is given by:

$$V = V_{\text{ref}} \left(\frac{d_1}{2} + \frac{d_2}{2^2} + \ldots + \frac{d_n}{2^n} \right)$$

V_{ref} is the full scale range of the output voltage. If the changes of the voltage V are plotted against the changes of the input binary word (Fig. 19) it can be seen that:

the output voltage is made up of a series of discrete voltages, each discrete value corresponding to a specific word;

the corresponding voltage points lie on a straight line which is called the ideal transfer characteristic;

when all the bits are in the state 1 the corresponding voltage differs from the reference voltage by one quantum:

$$V_{\text{max}} = \frac{V_{\text{ref}}}{2^n} (2^n - 1)$$

the error introduced is therefore dependent upon the number of bits of the DAC.

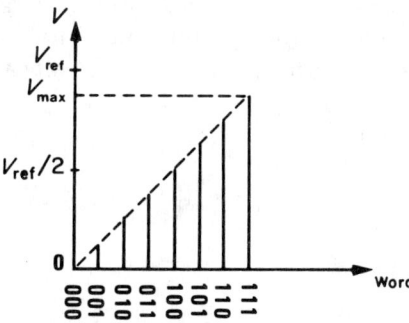

Figure 19

The second characteristic is the *amplitude of the output voltage*, V_{max}. It corresponds to the maximum variation of the output voltage when all the bits change from state 0 to state 1 and defines the dynamic range of the converter. This definition is equally valid for unipolar or bipolar converters though the expression for V_{max} depends upon the polarity of the DAC. Figure 20 shows the characteristics of a unipolar DAC and of the corresponding bipolar DAC with the same number of bits and the same amplitude of output voltage, but one has a reference voltage twice that of the other, so that in this case the quanta are equal.

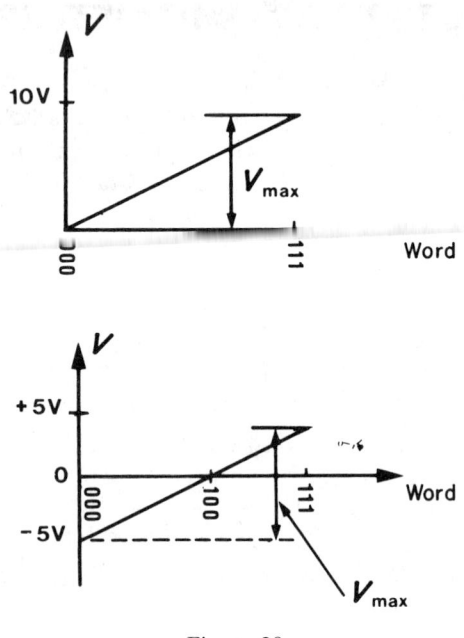

Figure 20

If a bipolar DAC has the same value of reference voltage as that of the unipolar DAC then the range of the output voltage is practically doubled. If the number of bits is kept the same the quantum is doubled and if the same value of quantum is desired then an extra bit has to be added to the word. The amplitude of the output voltage is sometimes called, wrongly, full scale. The amplitude of the output voltage can be expressed in terms of the reference voltage and of the number of bits thus:

for a *unipolar* converter

$$V_{max} = V_{ref}\left(1 - \frac{1}{2^n}\right)$$

and for a *bipolar* converter

$$V_{max} = V_{ref}\left(2 - \frac{1}{2^n}\right)$$

From these two expressions and the relation on p. 33 giving the value of the voltage V it is apparent that the output voltage can never attain the value V_{ref}. Attention must, therefore, be given to the required maximum value of the voltage to be obtained when deciding upon the length of word and the reference voltage to be used.

It should also be noted that the reference voltage and the quantum cannot both be expressed simultaneously as simple numbers. Thus for a quantum of 10 mV the reference voltage will have to be equal to 10.24 V or 5.120 V etc. depending on the value of n.

Next the *resolution* of a DAC must be defined. This characteristic is given by the number of bits that the converter can accept and it is an intrinsic theoretical datum. The resolution defines the amplitude of the smallest change of output voltage as a fraction of full scale, or taking into account the number of bits the input must handle, we can write:

$$\text{resolution} = \frac{\text{quantum}}{\text{full scale}} = \frac{1}{2^n}$$

that is

$$\text{resolution} \equiv \text{number of bits}.$$

The resolution can also be defined as the ratio of the smallest quantity that can be obtained (i.e. a quantum) to the maximum voltage that can effectively be obtained:

$$r = \frac{V_{ref}}{2^n} \cdot \frac{1}{V_{ref}\left(\dfrac{2^n - 1}{2^n}\right)} = \frac{1}{2^n - 1}$$

These two definitions are equivalent when n is sufficiently large, which is often the case. Table 3 gives the resolution calculated for various values of n.

Table 3. The resolution of different size DACs

n	Resolution	1 quantum as % V_{ref}
8	1/256	0.391
10	1/1024	0.0977
12	1/4096	0.0244
14	1/16384	0.0061

The *conversion time* allows us to calculate the speed of conversion of a DAC and hence obtain the maximum conversion frequency of the DAC, i.e. the maximum number of conversions that can be carried out per second. The conversion time is the time required for the output signal to reach the desired value within the specified error. It is dependent upon the components used, particularly the switches and amplifiers, and includes delays, different rise times, the damping of oscillations etc. The worst case is when all the bits change simultaneously from 0 to 1. Hence the conversion time is often defined as the time required for the output signal to sweep through the whole dynamic range of the converter, provided the result is obtained with an accuracy of $\pm\frac{1}{2}$ quantum (Fig. 21).

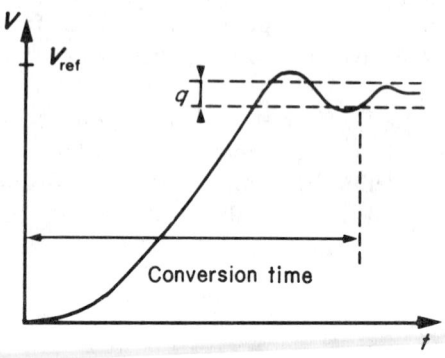

Figure 21

Manufacturers sometimes also specify the conversion time corresponding to a change of one quantum and the maximum possible rate of change of the output signal.

An important problem is that raised by *transient output conditions* called *glitches* which may arise when the digital input information gets altered. The switches controlled by this information have different switching times and this results in short-lived, but nevertheless unwelcome, erroneous digital words, which in turn result in unwanted output voltages. If for example a word must change from 011 to 100 (assuming $n = 3$) and if in so doing the MSB changes value before the other

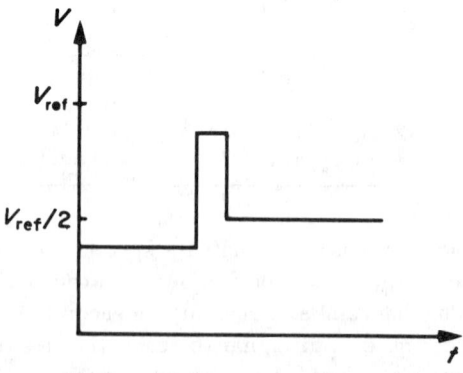

Figure 22

bits then an intermediate state 111 occurs. As long as that state persists the output voltage will be equal to $V_{ref}(1 - 1/2^n)$ (Fig. 22). The corresponding spike is the transient state or glitch. In practice this transient will be reduced by the finite pass band of the converter. There are two ways of eliminating these spikes:[21] a low-pass filter may be placed in the output which will smooth out the spikes (these spikes represent very high frequencies), or the output signal of the DAC can be sampled synchronously with the sampling of the input signal of the DAC but with a certain constant delay, thus allowing transient disturbances to die away during that waiting time.

The last important characteristic of a DAC is its *accuracy*. It is defined as the difference between the output signal obtained and the calculated theoretical value or the value based on the transfer function. The accuracy is usually expressed as a percentage of full scale, or as a fraction of the quantum and rarely in mV or in μV.

There are various sources of errors in DACs such as offset, gain factor, non linearity errors etc. which will be examined in detail further on. The accuracy quoted by manufacturers is usually $\pm\frac{1}{2}$ quantum. All bits are involved in the definition of the accuracy but their effect varies according to their weight. Indeed the MSB must be very accurate since it controls a corresponding voltage equal to half full scale with an absolute accuracy better than $\frac{1}{2}$ quantum, therefore its relative accuracy must be better than $1/2^n$. The LSB must also be accurate to $\frac{1}{2}$ quantum which corresponds to a relative accuracy of only $\frac{1}{2}$. This example has shown that great attention must be given at the design stage of a DAC to the bits of greatest weight, the other bits needing much less accuracy (which decreases further with their rank).

Table 4. A possible distribution of the error among different bits

Theoretical value of each bit	Absolute error admissible (mV)	Relative error for each bit ($\times 10^{-3}$)
5.00	20	4
2.50	20	8
1.25	10	8
0.625	10	16
0.3125	15	16
0.1562	15	32

As an example, Table 4 shows a possible distribution of the error among the different bits in order that the total sum of these errors be less than half a quantum. It is assumed that $n = 6$ and $V_{ref} = 10$ V. Hence the quantum has a value of 156 mV and the sum of the partial errors must be less than 78 mV. The total error in this case is equal to 70 mV, which is well below $q/2$.

2.4 ERRORS IN DACs

The performances given by a DAC generally differ from those predicted because of the actual errors that occur. The transfer characteristic has, for instance, the shape shown in Fig. 23 and when it is compared with the curve of Fig. 19 the following differences are apparent; the extreme points on the curve have values which differ from the corresponding values on the ideal curve; and the points in between these extreme values do not lie on a straight line. These differences are essentially due to: the *offset error*, the *gain error* and the *linearity error*.

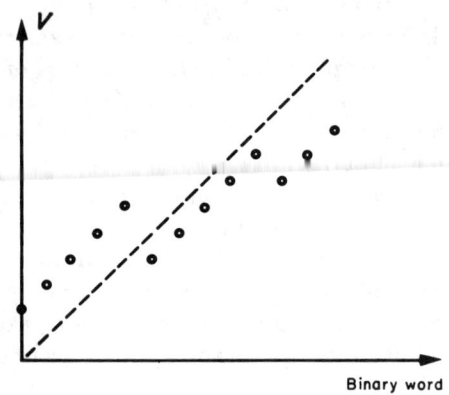

Figure 23

In this case the transfer function of a converter can be written:

$$V = (1+\Delta K)\,V_{\text{ref}}\left[d_1\left(\frac{1}{2}\pm W_1\right) + d_2\left(\frac{1}{2^2}\pm W_2\right) + \ldots + d_n\left(\frac{1}{2^n}\pm W_n\right)\right] + V_{\text{os}}$$

where V_{os} is the offset error, ΔK is the gain error and W_j is the weight error of the jth bit. The offset and gain errors are stated by the manufacturers whereas the weight error is not, because it can be dependent upon the state of other bits. This dependence of one bit upon another is called the *superposition error*. The weight error of a bit and the superposition error are combined into the linearity error. Each error is going to be defined on the assumption that no other errors are present though in practice several errors occur simultaneously and their effects are combined.

2.4.1 Offset Error

The offset error is the difference between the voltage output of the DAC when all the bits are in the state 0 and the voltage that ought to have been obtained. This error produces a *vertical shift* of the transfer characteristic (Fig. 24) and may be a

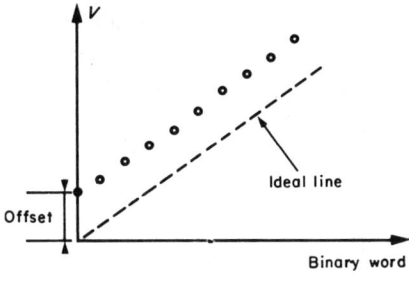

Figure 24

positive or negative shift. It is independent of the value of the input word and is usually expressed as a percentage of V_{ref}.

2.4.2 Gain Error

This error results in a *rotation* of the transfer characteristic about the point obtained when all the bits are in the state 0 (Fig. 25). In this case the value of the

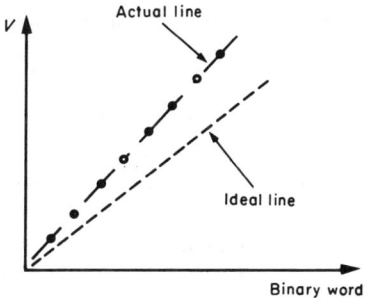

Figure 25

error is not constant but depends upon the input word, this error being the greatest when all the bits are in the state 1. The effects of gain and offset errors may be compensated for, at a given temperature, by adjusting the overall gain of the converter to 1 and adding a DC voltage to eliminate the offset error, though these errors reappear if the temperature changes.

2.4.3 Linearity Error

The conventional definition of linearity is as follows: a system is linear if the scatter between the transfer characteristic of the converter and the best straight line is less than $\frac{1}{2}$ a quantum. This implies that the transfer characteristic of the device must first be plotted in order to produce the best straight line that will show whether the DAC is linear. Clearly this definition is not very practical. The linearity error can also be defined as the difference between the output voltage obtained for a given word and the voltage, corresponding to the same word, which can be read off the straight line joining the two limiting voltages corresponding to the words 00 . . . 00 and 11 . . . 11 (Fig. 26). The difference ΔV represents the linearity error.

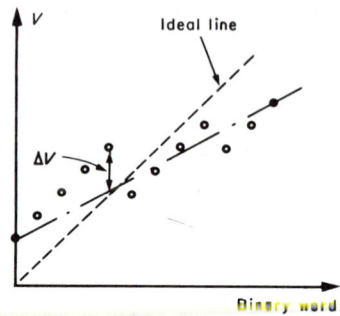

Figure 26

Finally, the linearity error is the *difference* between the actual output voltage and the corresponding voltage measured on the ideal transfer characteristic. These two last definitions are equivalent if the error is expressed as a relative error.

The *superposition error* defined previously may be better understood by stating that the whole is not equal to the sum of its parts. Having adjusted the converter so that the offset error is eliminated each bit is successively put into state 1, the other bits being held at 0, and the various voltages produced are summed. Then all the bits are put into state 1 at the same time and the voltage produced is compared with the previous result. If the two voltages differ then there is a superposition error and if the voltages are the same there is no error (this method may be used as a quick way of testing a converter).

2.4.4   Differential Linearity—Monotonicity

The differential linearity describes the behaviour of the output in relation to two consecutive input states. The *differential linearity error* is the difference between the value of the voltage change for two consecutive input states and one quantum. If this difference remains always equal to one quantum the differential linearity error is zero. A maximum differential linearity error of $+q/2$ (q being the quantum) means that the variation of the output signal may be within

$$\frac{q}{2}\left(q - \frac{q}{2}\right) \quad \text{and} \quad \frac{3}{2}q\left(q + \frac{q}{2}\right)$$

for two consecutive input states.

If the output voltage does not vary when the input state changes then the differential linearity error reaches one quantum. It can also happen that the output voltage diminishes when the binary number at the input increases and in this case the differential linearity error exceeds one quantum and the system is said to be non-linear. To say that a converter is *monotonic* amounts to saying that its output voltage increases or at least remains constant when the input number increases.

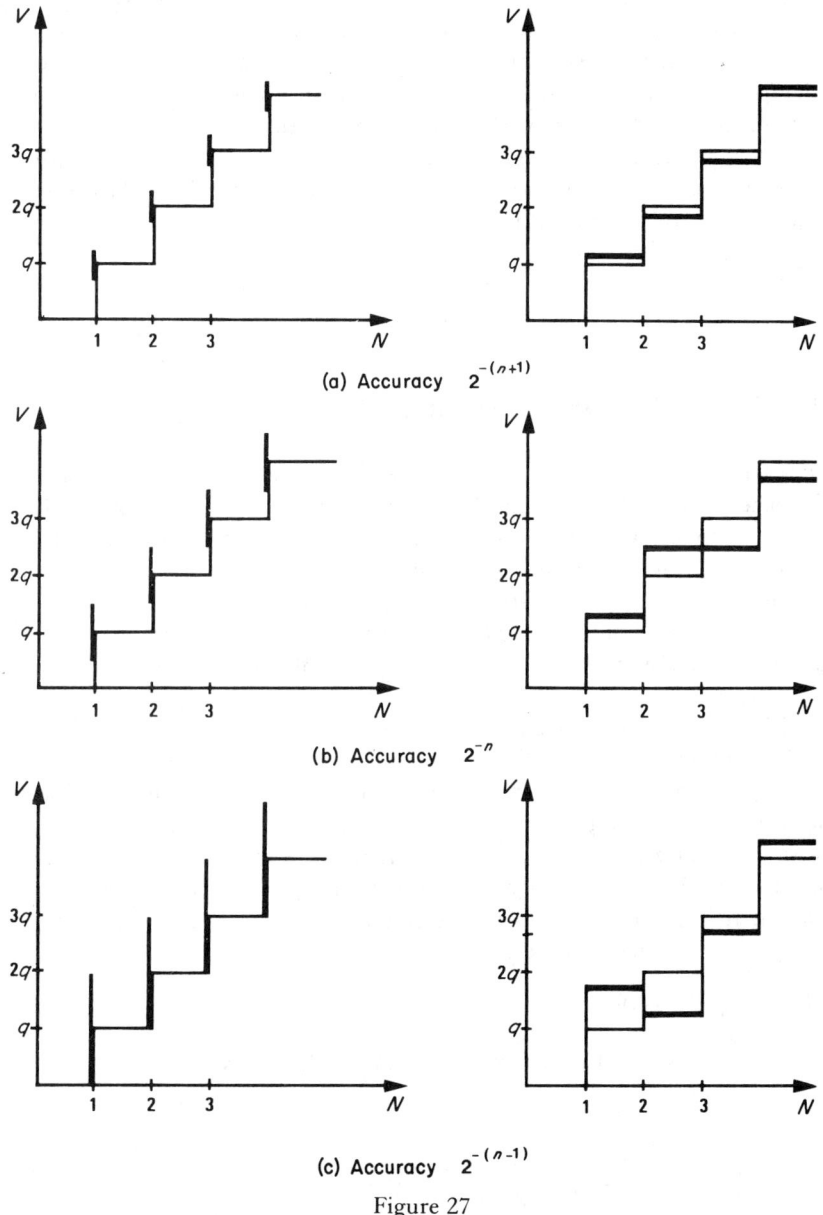

(a) Accuracy $2^{-(n+1)}$

(b) Accuracy 2^{-n}

(c) Accuracy $2^{-(n-1)}$

Figure 27

The ideas of resolution, accuracy and monotonicity can be shown to be related. Consider a word of n bits, the resolution is equal to $(1/2^n)$. If the accuracy is $2^{-(n+1)}$ the differential linearity error amounts at most to $q/2$ (that is $\pm q/4$). Figure 27(a) shows that when the transfer characteristic is followed faithfully a change in the input signal always brings about a like change of the output signal. If the accuracy amounts to 2^{-n} the linearity error is $\pm q/2$. In this case the output signal may remain constant if the input signal changes by a unit (Fig. 27(b)).

Finally if the accuracy is $2^{-(n-1)}$ the linearity error reaches $\pm q$ and the output signal may decrease while the input signal increases (Fig. 30(c)) and the system is not monotonic. Therefore a converter is monotonic when the accuracy is better or at least equal to the resolution. It is however sometimes useful to have a resolution better than the accuracy, for example 2^{-12} and 2^{-10} respectively. For large signals the accuracy will be 2^{-10}, as given by the theory, but for weak signals (or for small changes) the accuracy will be equal to the resolution, that is 2^{-12}. If the converter must handle low values it would therefore be useful for it to have a high resolution even if the accuracy is lower.

The differential linearity is a necessary but not sufficient condition for linearity. Moreover a linearity of $\pm q/2$ guarantees a differential linearity of $\pm q$.

2.4.5 The Influence of Temperature

One must not underestimate the influence of temperature on the operation of a DAC, for it can be at the root of faulty or even inconsistent operation. When the temperature changes it brings about changes in the offset, gain and linearity errors. These changes in the errors are added on to those errors already present at the initial temperature. They are usually specified in % °C^{-1}. Knowledge of the operating temperature range of the converter enables the additional errors due to temperature changes to be calculated. An example will illustrate the way this problem is tackled.

Consider a converter with the following specifications: resolution, 12 bits; full scale, 20 V (1 quantum = 4.88 mV); variation in offset error, ± 10 ppm °C^{-1}; variation in gain error, ± 20 ppm °C^{-1}; sensitivity of $+15$ V supply, $\pm 0.02\%$ per % variation of supply voltage and $\pm 0.018\%$ °C^{-1}; and linearity error, $\pm q/2$.

Assuming that the temperature may vary by 30 °C, what is the resulting maximum error? The various errors and the changes in these errors must be calculated. It is assumed that both the gain and offset errors had been compensated for at the ambient temperature.

Variation in offset error:

$$\Delta V_1 = (10 \times 10^{-6}) \times (20 \times 30) = 6 \text{ mV}$$

that is

$$\Delta V_1 = \frac{6}{4.88} = \pm 1.23q$$

Variation in gain error:

$$\Delta V_2 = (20 \times 10^{-6}) \times (20 \times 30) = 12 \text{ mV}$$

that is

$$\Delta V_2 = \frac{12}{4.88} = \pm 2.46q$$

Effect of supply sensitivity:

$$\Delta V_3 = \frac{(2 \times 10^{-4}) \times 20 \times (1.8 \times 10^{-4}) \times 30}{0.01}$$

$$= 2.16 \text{ mV}$$

that is

$$\Delta V_3 = \frac{2.16}{4.88} = \pm 0.44q$$

The total error is therefore the sum of these three terms plus the linearity error which is independent of the temperature, thus:

$$\varepsilon = \pm(2.46 + 1.23 + 0.44 + 0.5)q = \pm 4.63q$$

This error is about eight times greater than the linearity error. If the temperature rises by 30 °C the resolution will therefore be divided by eight and to all intents and purposes the converter would appear to have only nine significant bits.

A chart supplied by the firm Burn–Brown[22] and reproduced in Fig. 28 enables one to find out very rapidly the operating temperature range within which a converter will have a variation of error of less than one quantum, also taking into account the number of bits and the manufacturer's specifications. Referring to the previous example it can be seen that the maximum permissible temperature variation is 10 °C.

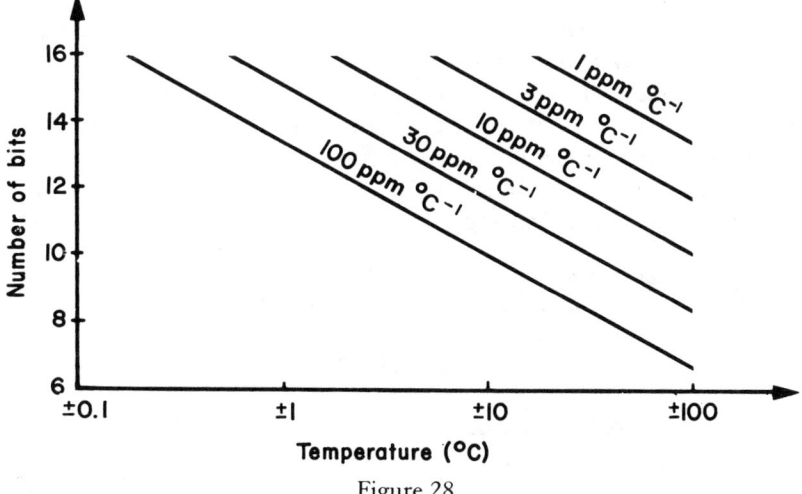

Figure 28

2.5 ANALYSIS OF THE MAIN DACs

Having defined the characteristics and the errors of DACs the analysis of the principles of operation of the most commonly used DACs will now be undertaken followed by a few practical examples. DACs are grouped into two families according to whether the binary word is converted directly into an analog signal, *direct DAC*, or if an intermediate parameter is used, *indirect DAC*. Moreover direct DACs are subdivided into *parallel* DACs and *sequential* or *serial* DACs according to whether the conversion of the various bits is made simultaneously or in sequence.

2.5.1 Parallel DACs

The basic schematic diagram of a parallel DAC may be deduced directly from its transfer function:

$$V = d_1 \frac{V_{ref}}{2} + d_2 \frac{V_{ref}}{4} + \ldots + d_n \frac{V_{ref}}{2^n}$$

It has the following elements: a *reference quantity*; a *weighting system*, multiplication by the coefficient of the binary series $1/2, 1/4, \ldots, 1/2^n$; a *digital control*, multiplication by d_i which is 0 or 1; a *summation* of these various signals (voltages or currents); and possibly a *transformation* $V \rightarrow I$ or $I \rightarrow V$.

It is usually easier to add currents than voltages and therefore the defining equation is modified to make it more suitable for that operation and becomes:

$$V = R\left(d_1 \frac{V_{ref}}{2R} + d_2 \frac{V_{ref}}{4R} + \ldots + d_n \frac{V_{ref}}{2^n R}\right)$$

$$V = R(d_1 I_1 + d_2 I_2 + \ldots + d_n I_n)$$

After having summed the weighted currents, the current to voltage transformation is carried out, using for instance the circuit in Fig. 29. The amplifier used must be chosen with care if it is not to affect the performance of the DAC. Other work should be consulted in order to define the criteria for its selection.[23,24]

The accuracy and speed of these converters depend essentially upon the components used and their effects will be discussed in another part of the book. The functional schematic drawing of a parallel DAC is shown in Fig. 30. The

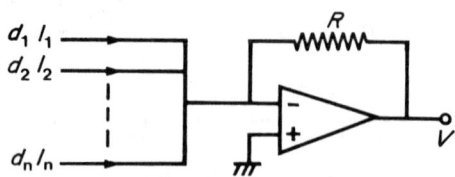

Figure 29

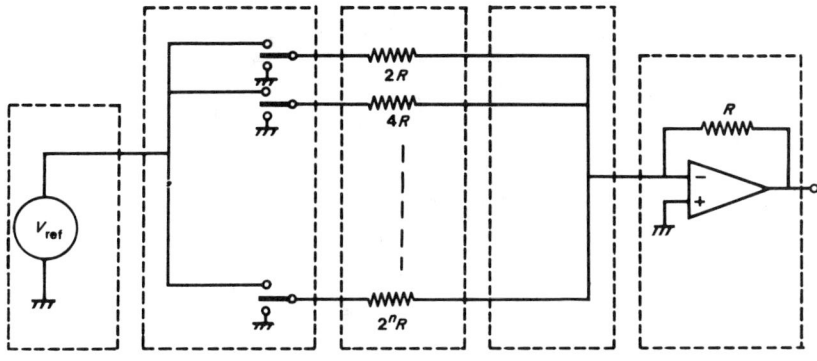

Figure 30

weighting used is a V/I weighting for it transforms the voltage V_{ref} into a current

$$I_i = \frac{V_{ref}}{2^i R}$$

This system has many alternatives; one can for example carry out first the weighting and then the current control. The switches which carry out a control (or command) function are themselves controlled by the different bits of the binary word to be decoded. According to whether this bit is in the state 0 or 1, the current passing through the corresponding resistor may or may not be fed to the amplifier to be added to the other currents.

These converters are very fast since the information corresponding to the binary word is instantly available. It is possible to do away with the output amplifier and use the current delivered by the converter, thus increasing the conversion speed. The three essential components of this system are then the reference source, the network of switches and the weighting network. The DACs currently manufactured differ mainly in the way the weighting is done, the solution adopted often being dictated by the manufacturing techniques available.

In order to analyse the different types of DACs it will be assumed to start with that the code used is the natural binary code, corresponding to unipolar signals (usually positive). Later it will be shown how it is possible to adapt them for bipolar signals and the other codes.

2.5.1.1 Weighted resistor converter

This is the simplest possible converter and makes use of the basic circuit diagram of Fig. 30. By means of resistors $2R$, $4R$, $8R$, ... $2^n R$, weighted currents are obtained which are summed using an operational amplifier (Fig. 31). Thus the output signal is proportional to the binary word to be converted. A network of switches makes it possible to sum only the currents required. The output voltage obtained is given by:

$$V = -R \sum_{i=0}^{n-1} d_i I_i$$

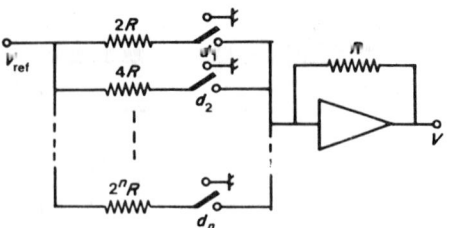

Figure 31

with

$$I_i = \frac{V_{\text{ref}}}{2^i R}$$

The minus sign is introduced by the summing amplifier. It is possible to interchange the position of the resistors and switches as in Figs. 30 or 31 but the connection of Fig. 31 is however more useful since the switches controlled by the coefficients d_i then direct the current $d_i I_i$ either to the amplifier or to earth so that current always flows through the resistor $2^i R$. This reduces the conversion time and provides the source V_{ref} with a constant load.

The attraction of such a converter lies in its *simplicity* but it fairly soon becomes inaccurate as the number of bits is increased. The sources of error are mainly the *switches*, the *weighting resistors* and the *output amplifier*. When switches are used as in Fig. 30, the currents through the switches change in the ratio 1 to 2^{n-1} and can be large compared with the allowable through current of the switch (particularly if the switch needs to be fast) but small compared to its leakage current. Moreover in the ON state the switches introduce a voltage error (if junction transistors are used), or introduce a resistance which is not negligible (if FETs are used).

The weighting resistors must be in the ratio of 1 to 2^{n-1}. It is difficult in practice to obtain resistances of such a wide range of values and yet of the same tolerance and, above all, the same temperature coefficient. Moreover these resistors are difficult to manufacture in microelectronic circuits. These difficulties therefore have led to the search for other solutions. A simple way of resolving this problem is to use several identical blocks each of four current generators of the above type; these blocks are sometimes called *quads*. The output signal from these various blocks, be it voltage or current, will be weighted in accordance with the code used and their position in the array of blocks used. This is actually one of the most commonly used methods for producing a 10 or 12 bit DAC as the various elements can be manufactured as integrated components.

Figure 32 shows a 12 bit DAC using this principle. The three currents I_a, I_b, I_c have identical expressions (to the nearest bit number) which is ensured by having identical sets of resistors and switches. These currents must be weighted, prior to being added together, by the factors 1, 1/16 and 1/256 which is done by a second set of resistors. However this method requires some care in the choice of components used, and in particular the resistors used for the second weighting

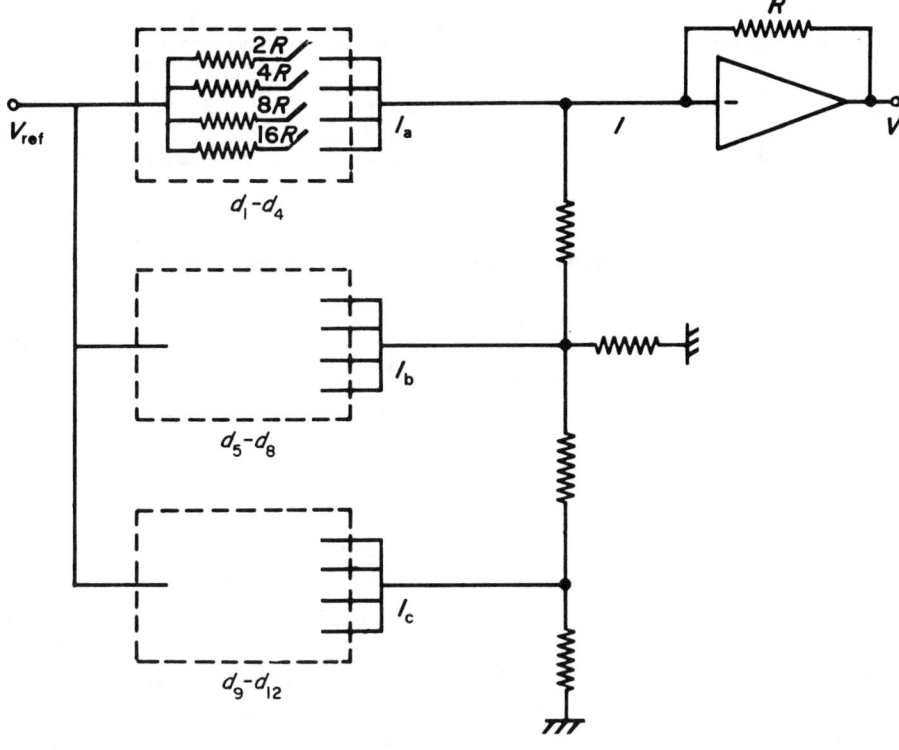

Figure 32

must be current and not voltage fed. The output signal is:

$$V = R\left(d_1\frac{V_{\text{ref}}}{2R}+\ldots+d_4\frac{V_{\text{ref}}}{16R}\right)+\frac{1}{16}\left(d_5\frac{V_{\text{ref}}}{2R}+\ldots+d_8\frac{V_{\text{ref}}}{16R}\right)$$
$$+\frac{1}{256}\left(d_9\frac{V_{\text{ref}}}{2R}+\ldots+d_{12}\frac{V_{\text{ref}}}{16R}\right)$$

or

$$V = R(I_a+\tfrac{1}{16}I_b+\tfrac{1}{256}I_c)$$

and it can be seen that each bit has in fact the weight corresponding to its rank in the expression for V. This method is very useful in the case of the BCD code for it is only necessary to alter the resistance values of the second network in order to get a power of ten relationship for the currents I_a, I_b and I_c.

2.5.1.2 The ladder converter

The arrangement of Fig. 32 uses only four resistance values for the basic circuits, which can easily be obtained in practice. It is possible to reduce this number further, to two distinct values of resistor R and $2R$ by using a ladder network. The principle of the method is shown in the diagrams of Fig. 33. In Fig. 33(a) each

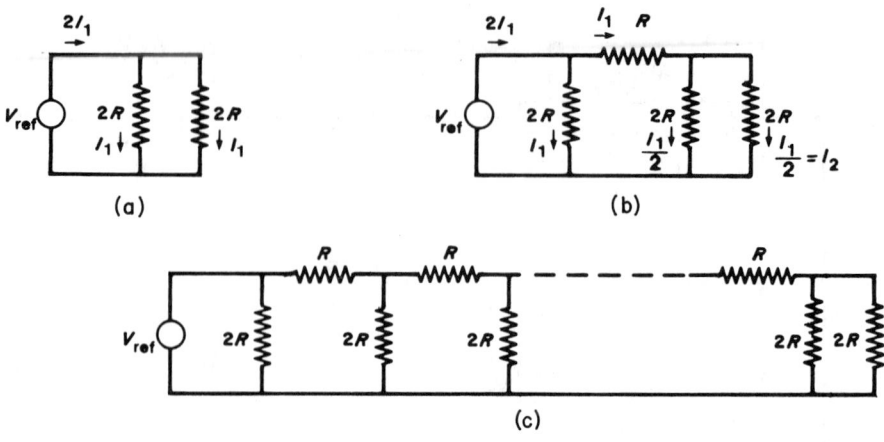

Figure 33

resistor $2R$ has a current $I_1 = V_{ref}/2R$ flowing through it, the source supplying a current $2I_1 = V_{ref}/R$. One of these two resistors can be replaced by two resistors R in series, (Fig. 36(b)) having current I_1 through them. Again one of those two resistors can be replaced by two resistors $2R$ in parallel, having a current $I_2 = I_1/2 = V_{ref}/4R$ flowing through them. Current I_1 has therefore been divided by 2. It is possible to repeat this process n times and the diagram of Fig. 36(c) is obtained, in which all the resistors having the value $2R$ have currents flowing through them in a geometric progression of common ratio $\frac{1}{2}$. This network is the basic element of ladder converters. Figure 34 illustrates one version of a ladder

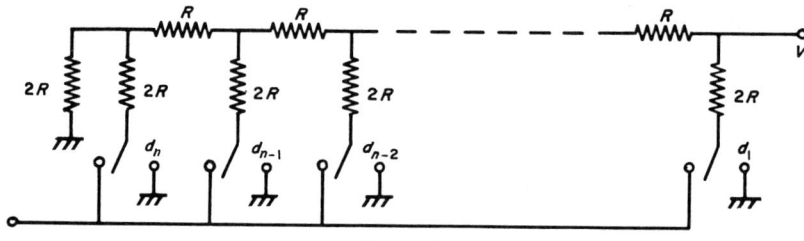

Figure 34

converter. Each switch can connect the corresponding resistor $2R$ either to the reference voltage source V_{ref} or to earth. The switch corresponding to the LSB is on the left and the switch corresponding to the MSB on the right. The value of the resistance seen at the output is R whatever the switch positions. Using the results obtained from Fig. 33 and Thevenin's theorem it can easily be shown that if the weighted bit i is 1, all the other bits being 0, then the output voltage is given by $V_{ref}/2^i$.

Hence, by the superposition theorem, the voltage for an n bit word is

$$V = V_{ref}\left(\frac{d_1}{2} + \frac{d_2}{4} + \ldots + \frac{d_n}{2^n}\right)$$

The two important components which determine the accuracy of the converter are the resistor network and the group of switches. The switches have a resistance which is in series with the resistors $2R$ of the network. Since the value of that resistance depends upon the current through the switch, the accuracy of the system could be reduced. As regards the ladder resistors their absolute value is of little importance since only their ratio is involved but a very good relative accuracy is required together with a low temperature coefficient. However this type of converter has certain disadvantages:

relatively large voltages must be switched (V_{ref} is usually of the order of 10 V),
the currents circulating through the resistors $2R$ change direction during switching resulting in transient conditions,
to achieve a high switching speed, low resistance values must be used in order to reduce the time constants, but then in this case the parasitic resistance values due to the switches cannot be neglected.

These disadvantages can be partly reduced by using an *inverted ladder converter* in which currents and no longer voltages are switched (Fig. 35). In this connection

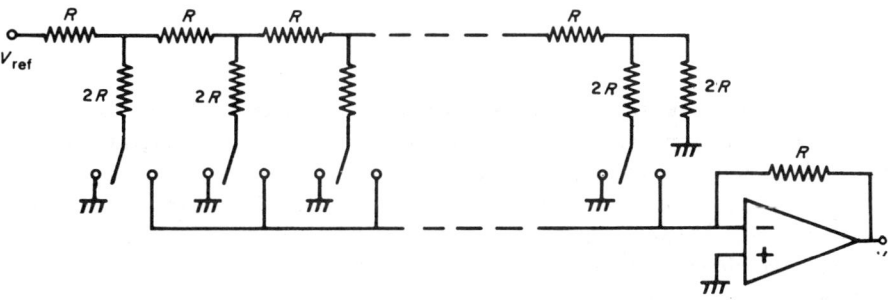

Figure 35

diagram, the MSB is on the left and the LSB on the right. The currents through the resistors $2R$ always flow in the same direction. According to whether the bit is in state 1 or state 0, this current flows towards the summing amplifier or flows to earth. Whatever the switch position the end of the resistors is always earthed since the amplifier input is at a virtual earth. The current through these resistors is therefore constant and it is possible to increase the network impedance and reduce the influence of parasitic elements on the dynamic behaviour of the converter.

2.5.1.3 DAC using bipolar codes

The codes most used in practice are bipolar codes and it is useful to see how the previous circuit arrangements can be modified to make them suitable for bipolar codes.

The most important codes for D–A conversion are the offset binary code and two's complement code which only differ by the MSB (sign bit). If a weighted resistor converter such as that shown in Fig. 31 is to be used, it is only necessary to add a resistor equal to that of the greatest weight, i.e. $2R$, and to connect it to $-V_{ref}$ (Fig. 36). This will provide a permanent current of $-V_{ref}/2R$ and the expression for the output signal becomes:

$$V = -V_{ref}\left(\frac{d_1}{2} + \frac{d_2}{2^2} + \ldots + \frac{d_n}{2^n}\right) + \frac{V_{ref}}{2}$$

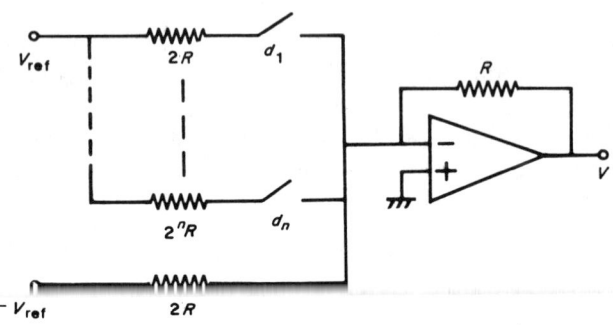

Figure 36

One must not forget that in this case the dynamic range of the output signal is unchanged but simply offset by the quantity $-V_{ref}/2$ (and so centered about zero). If it is required to have a dynamic range from $-V_{ref}$ to $+V_{ref}$ then the amplifier gain has to be doubled. To achieve this it is only necessary to use a feedback resistor having a value of $2R$. In the case of a ladder converter the ends of $2R$ resistors which are earthed (because the corresponding bit has value 0) must be connected to a reference voltage of $-V_{ref}$. The *magnitude sign code* enables the amplitude and sign of a signal to be coded separately, so in order to convert a number represented in this code, the MSB is separated from the other bits. The remaining $(n-1)$ bits are fed to a unipolar DAC which produces a current I proportional to the word corresponding to these $(n-1)$ bits. The state of the MSB is used to select one of the voltages $-RI$ and $+RI$ derived from the current I, the selection being made by means of a switch controlled by this bit (Fig. 37).

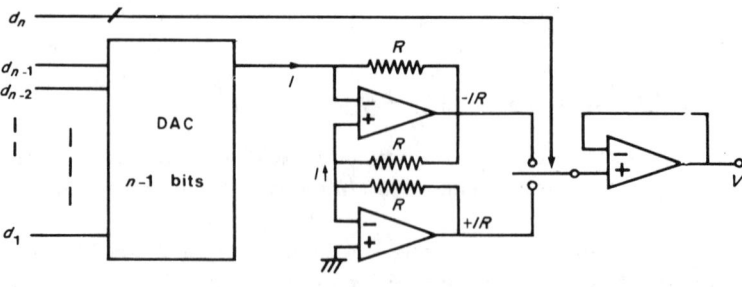

Figure 37

2.5.1.4 Example of a commercial DAC

To end this part of the chapter concerned with parallel type DACs, an example of a commercial DAC[25,26] using weighted resistors will now be described in detail. In a later chapter the various methods of manufacturing the main component parts of a DAC such as the voltage source, resistance network and switches, will be reviewed in order to compare their complexity, the possibility of integration and above all their accuracy. Figure 38 shows the schematic diagram of a

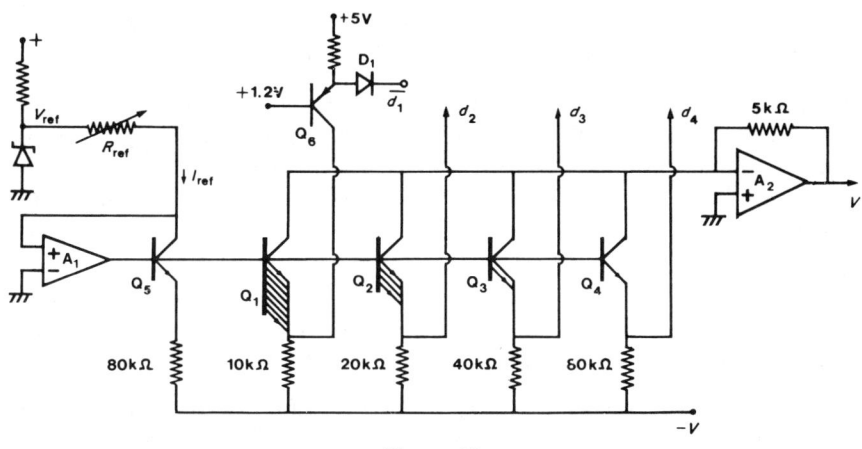

Figure 38

four bit converter as currently manufactured. Usually transistors Q_1 to Q_4 and their driving circuits are grouped into one integrated circuit and the reference source and the weighted resistors form two other sub-assemblies. Resistors of 10 kΩ to 80 kΩ are used for current weighting, the currents vary from 1 mA to 0.125 mA. These currents are fed to the output amplifier through four transistors Q_1 to Q_4 in the common base configuration. These transistors are controlled by four switching circuits, only one of which is shown in the diagram, consisting of a transistor and a diode which receives one of the bits of the word to be converted. The output amplifier A_2 gives a voltage V derived from the summation of the weighted currents. The reference voltage is obtained by the arrangement of the Zener diode, resistor R_{ref} and amplifier A_1. Transistor Q_5, which is identical to transistor Q_4, supplies the reference current for the automatic adjustment arrangement of the base voltage common to the four transistors Q_1 to Q_4, controlled by this reference.

First, the operation of the source supplying the reference voltage will be examined. The voltage at the terminals of the temperature compensated Zener diode is V_{ref}. The voltage across resistor R_{ref} which is earthed through the amplifier is V_{ref}, hence a current $I_{ref} = V_{ref}/R_{ref}$ flows through it. This current can be adjusted to the required value by adjusting R_{ref}. The current thus produced flows to the collector of transistor Q_5 and, to the extent that the current gain β of Q_5 is much greater than one, the current is equal to the emitter current which flows through the 80 kΩ resistor. The supply voltage $-V$ is fixed and so is the emitter voltage of Q_5, and the base voltage is also the same to within 0.6 V. Thus with a voltage $-V = -15$ volts and a current $I_{ref} = 0.125$ mA the emitter voltage of Q_5 is equal to -5 V and that of the base to about -4.4 V. If the temperature, or one of the transistor parameters changes, the voltage of the base will change, but the emitter current will remain practically constant since it is supplied by a circuit which is independent of transistor Q_5. The arrangement Q_6, D_1 represents the first current commutation controlled by the binary parameter d_1. The base of transistor Q_6 is raised to a voltage of 1.2 V which closely corresponds to half the sum of the voltages associated with the two TTL logic states 0 and 1.

The operation of this circuit arrangement is as follows:

if $d_1 = 1$, $\bar{d}_1 = 0$, a zero voltage is applied to diode D_1 which conducts and draws all the current available. Transistor Q_6 is then off. Transistor Q_1 is on and a current of 1 mA passes through it and the output circuit.

if $d_1 = 0$, $\bar{d}_1 = 1$, a voltage exceeding 2 V is applied on the cathode of D_1 which cuts off and transistor Q_6 can conduct. The resistor of its emitter is so chosen that the current flowing through it will be greater than 1 mA. This current then flows through the 10 kΩ weighting resistor and the emitter of Q_1 is raised to a voltage greater than that of its base (and fixed by transistor Q_5) hence Q_1 cuts off and only its leakage current flows through it (a few nA).

The function of transistors Q_1 to Q_4 is the most important factor affecting the accuracy of the converter. If transistors Q_3 and Q_4 were identical and had unequal currents flowing through them in the ratio 1 : 2 (0.25 mA and 0.125 mA respectively) then their V_{BE} voltages would not be equal and since they are in common base mode they would introduce an error in the ratio of their emitter currents (a voltage difference of about 20 mV between the voltages at the 40 kΩ and 80 kΩ resistor terminals). In order to eliminate this error transistors Q_1 to Q_4 are not chosen identical but are made with emitter surface areas in a geometric progression of common ratio $\frac{1}{2}$. Since Q_1 to Q_4 carry currents obeying the same progression their V_{BE} voltages will always be equal, and since their bases are common their emitter voltages are also equal and in this way the required current accuracy is achieved. Transistor Q_5 adjusts automatically the base voltage of transistor Q_1 to Q_4 for any temperature variation that may occur. Current I_{ref} being fixed and assumed independent of the temperature the emitter voltage of Q_5 remains constant. If the temperature changes its base voltage varies and therefore the base voltages of transistor Q_1 to Q_4 also vary, but since these transistors are fabricated on the same chip they are well matched and their voltage V_{BE} will change in the same manner as that of Q_5. Thus their emitters will always remain at the same potential and the accuracy will be maintained provided the resistors are not affected by the temperature variations (or vary in identical fashion).

In order to increase the resolution of the converter it is only necessary to add other modules identical to the module which carries out the current weighting, and then to obtain the weighted sum of their currents. Monolithic 10 or 12 bit DACs are now available which require no external components. This miniaturization, and superior performance, has been made possible only by circuit integration, all the components being made with semiconductors.

Nowadays 10 bit DACs are available using complementary MOS techniques[27] and hence with a low power dissipation (typically 20 mW) thus avoiding quite a few of the usual problems. The advantages of this technique can be best appreciated by examining the limitations due to conventional techniques, with which it is difficult to obtain an accuracy greater than ten bits with good performance (at low cost), because the switches have finite current gains and this necessitates identical voltages, V_{BE} for all the transistors, and resistive networks with calibrated resistors, while the internal power dissipation limits the performance by producing thermal gradients.

Complementary MOS (CMOS) devices avoid these drawbacks. They have nearly infinite current gain. In an MOS circuit there is no equivalent to the voltage drop V_{BE} which occurs in a bipolar transistor. Instead a CMOS switch which is on is equivalent to a resistor whose value can be adjusted by controlling the geometry of the device. In a DAC made using CMOS techniques there is no diffused resistor, thus avoiding temperature variation problems. The R–$2R$ network is made up of thin film resistors deposited on the substrate.

In the chapter devoted to the fabrication of the various components, a DAC made using CMOS techniques will be described in detail in order to show how this technology can solve the problems of accuracy.

2.5.1.5 Multiplier DACs

The transfer function of the DACs just examined can be written as:

$$V = N \frac{V_{ref}}{2^n}$$

the voltage V_{ref} being assumed constant. When the reference voltage varies the DAC is called a *multiplying converter* since at every instant of time it gives the product of two variable quantities, one of which is an analog quantity and the other a digital quantity. In this case the reference voltage must be external to the converter and its characteristics affect the output signal; its most important characteristic is its maximum frequency.

The various techniques discussed previously are still applicable in so far as the respective components can operate with a variable reference voltage. The switches will be the most critical components, particularly if the voltage V_{ref} can be alternately positive and negative. When the reference voltage may change sign and bipolar codes are used there are four possible combinations for the sign of the output voltage. If the various possible values of V_{ref} and of N are plotted on two orthogonal axes then each sign combination corresponds to one of the quadrants. When the result can lie in any one of the four quadrants then the converters are referred to as *four quadrant multipliers*.

2.5.1.6 The advantages of an input interface

Up to now the way in which digital signals are presented to the input of the converter has not been discussed. It was assumed that the signals were there when a conversion operation was to be carried out, but parallel type DACs will allow signals at the input at all instants of time. Each change in a bit will be converted immediately, appearing as a new value of the output voltage, the only limitation being the switching time.

On the other hand, all the circuits following the converter should only receive a signal corresponding to particular input messages even though the bits are not all input at the same time. Similarly digital control circuits often supply output information at specific instants of time only. Hence isolating or interface circuits must be provided which store the digital input signals during the waiting period. These signals can be serial or parallel and therefore two solutions must be provided.

When the signals to be dealt with are *parallel*, the interface circuit consists of an arrangement of n latches, one for each bit as shown on the schematic drawing of Fig. 39(a). The input signals are only taken into account when the control signal is on. If the incoming signals are serial, the isolation circuit can still comprise n latches (Fig. 39(b)). These latches are connected so as to form a shift register. A latching pulse train, synchronous with the serial word must be provided, where

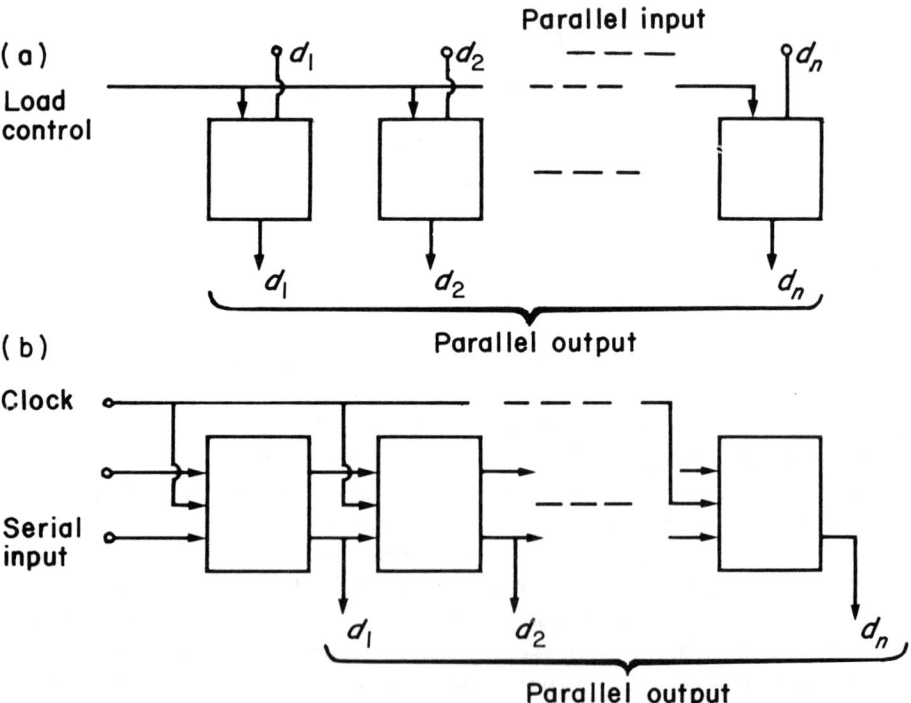

Figure 39

each shifting pulse allows the transfer of data from one latch to the next. To shift an n bit word therefore requires n pulses and n clock periods. The output signal is available in parallel form as soon as the shifting operation is completed. During the time of this operation the output signal of the DAC has no meaning since it does not correspond to a word normally present at the input. To overcome this drawback the circuit arrangement of Fig. 39(b) may be used as an input to the circuit of Fig. 39(a), the ON control signal only being sent at the end of the shifting operation.

2.5.2 Serial DACs

The parallel digital to analog converters described above are usually very fast, but require many precision components. Moreover if the word is available in serial form then the possibilities offered by these converters cannot be used to the best advantage as has been seen. The *serial DACs* have been conceived to overcome these drawbacks.

In a parallel converter each state can at any instant of time be changed independently of all the other states, since they are all acting simultaneously. In a serial converter (also called sequential converter) the operations are no longer simultaneous. Each bit is treated separately and its effect influences the signal generated by the following bits. If one of the bits of the input signal changes, the whole conversion process must be started again. In these converters the input signal must be available in serial form, with the LSB usually at the start and a clock is necessary to regulate the rate of the operations. The basic schematic circuit diagram for these converters is shown in Fig. 40. The signal to be converted is fed

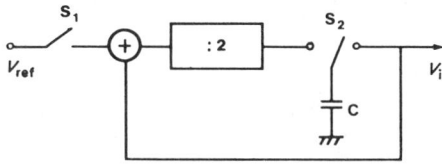

Figure 40

to the input synchronously with the clock signals and controls the operation of the converter bit by bit. The change-over switch S_2 is operated by the clock signal and switch S_1 by the binary signal that is to be decoded. During each clock period the change-over switch S_2 goes from left to right and vice versa, whereas switch S_1 closes if the bit at the input at that instant of time is 1. The principle of operation is as follows:

if the bit at the input is 1 the voltage V_{ref} is added to the voltage V_i reached at the end of the previous period and stored in a capacitor, the sum is divided by two and the resultant voltage in its turn is stored.

if the bit at the input is 0, only voltage V_i is divided by two and stored.

The voltage V_{i+1} obtained through this operation is stored and will be used during the next period. Thus one can write:

$$V_{i+1} = \tfrac{1}{2}(V_i + d_{i+1} V_{\text{ref}})$$

d_{i+1} being the value of the input bit, 0 or 1. Figure 41 shows the timing diagram for such a converter in the case of a six bit word expressed as 101001 (or 41 in decimal). It must not be forgotten that the LSB must be fed first. The signal V_i obtained during the last clock pulse is the output signal of the converter, and has the correct value for a short period of time only. A sample and hold circuit must be added after the DAC in order to obtain a continuous output voltage.

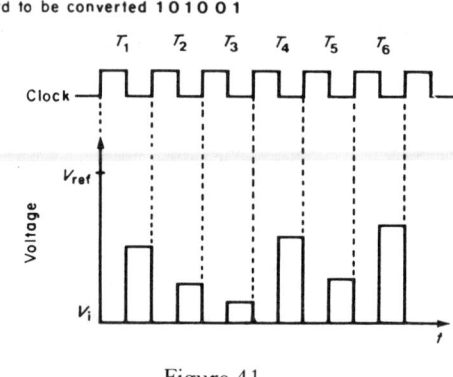

Figure 41

One of the advantages of this type of converter is that the complexity of the arrangement does not depend upon the number of bits. On the other hand the number of bits affects the conversion time since one clock period is needed per bit. The implementation of the schematic diagram of Fig. 40 actually requires a large number of operational amplifiers in order to store the various voltages and therefore other less complex but similar devices will preferably be used in practice.

One example of interest is the *serial hold converter*, which is shown in Fig. 42(a), and the timing diagram for a six bit word is given in Fig. 42(b). The conversion operation requires as many clock periods as there are bits in the word to be converted (in this case six). After these n periods the system delivers an output signal S, which lasts a clock period and allows a sample of the voltage V to be extracted. During the first half of a clock period T_i, S_2 is closed and the first amplifier A_1 sums the voltage $d_i V_{\text{ref}}$ and the output voltage V_{A2} of the second amplifier, and this sum is scaled by the coefficient $\tfrac{1}{2}$ due to the value of the resistors. Thus:

$$V_{A1} = -\tfrac{1}{2}(d_i V_{\text{ref}} + V_{A2})$$

d_i being equal to 0 or 1 according to the state of the input bit during this period. The binary word must be available in serial form with the LSB being first.

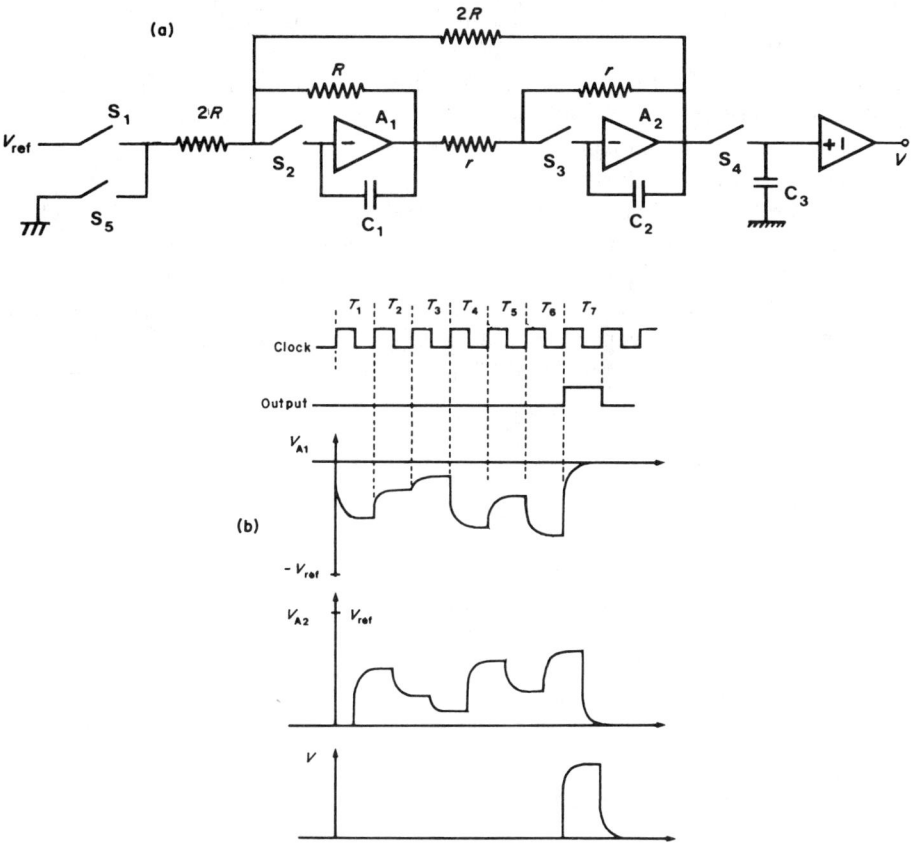

Figure 42

In the example chosen d_n is 1 during T_1, the output V_{A2} is zero and hence:

$$V_{A1} = -\tfrac{1}{2}V_{ref}$$

During the second half of the period T_1, S_2 is open and the capacitor holds the output voltage of A_1 constant. S_3 is closed and amplifier A_2 inverts the output voltage of A_1. The process is repeated until the appearance of the last pulse, called the synchronization pulse, T_7. During the first half of T_7, S_4 is closed and the output voltage of A_2 is transferred onto the capacitor C_3 and thus the output voltage V is obtained. Moreover S_5 and S_2 are closed thus discharging C_1. During the second half of T_7 capacitor C_2 is discharged, by connecting it to the output voltage of A_1 which is zero.

In order to obtain an accurate result the half period of the clock must be greater than the charging time constant of the capacitors. The ratio between these two quantities depends upon the accuracy required, that is upon the number of bits of the word to be converted (a ratio of $1:4.6$ is required for a 1% accuracy). The absolute value of the capacitances need not be known with great accuracy and it is only necessary that their values do not change during the time taken by one conversion and that the value be sufficiently large for the offset current of the amplifiers not to affect the output voltage during the conversion time. The only precision components required are the resistors R and r which determine the accuracy of the summation and of the division.

A second example is that of the *capacitive charge redistribution converter* which uses the idea behind the Shannon–Rack converter. During the first half of a clock pulse, at the instant of time when the bit present has the value 1, the reference voltage V_{ref} charges a capacitor C (Fig. 43). If this bit is 0 this capacitor is short-circuited. During the second half of the clock period the charge on C is partly redistributed on capacitor C'. If C = C' both charges are equal to half the sum of the initial charges on C and C', hence the operation carried out gives:

$$V_{i+1} = \tfrac{1}{2}(V_i + d_i V_{ref})$$

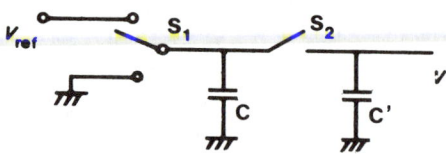

Figure 43

2.5.3 Indirect Converters

The second family of DACs covers the indirect converters, those converters which make use of an *intermediate parameter* during the conversion of the input binary word into the analog voltage output. This intermediate signal can be analog or digital. Such a procedure reduces the number of precision components needed but at the expense of the conversion speed which is reduced. The design of such devices led to the adoption of sequential methods which resulted in devices where logic circuits play a major role, and it is for these reasons that these A–D conversion methods are sometimes called sequential methods. Thus simple, low cost monolithic DACs will be produced, but their conversion speed will not be very high because the output signal must be extracted by averaging.

2.5.3.1 The intermediate parameter DAC using pulses

The first type of indirect DAC is one in which the word to be decoded is transformed into a pulse whose width is proportional to the word or into a sequence of pulses (or pulse train) in which the number of pulses is proportional to the word. These devices can be seen as consisting of two parts; a digital part and an analog part (Fig. 44).

The digital part converts the binary word into a pulse for example, by means of purely digital circuits (or logic). The second part transforms this intermediate

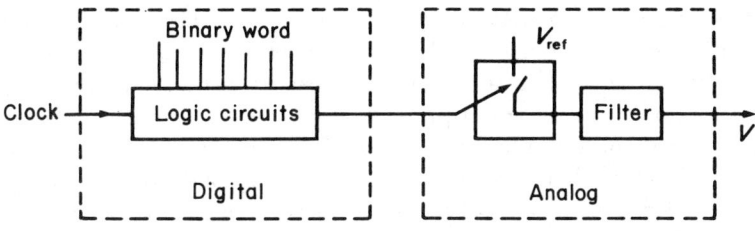

Figure 44

signal into a continuous voltage by means of an analog switch and a low-pass filter. Figure 45 shows the schematic diagram of an indirect DAC using pulse width modulation. It consists essentially of a buffer register which allows the binary word to be constantly available, an n bit down couter, an analog switch switching on or off the voltage V_{ref} at the output and a low-pass output filter.

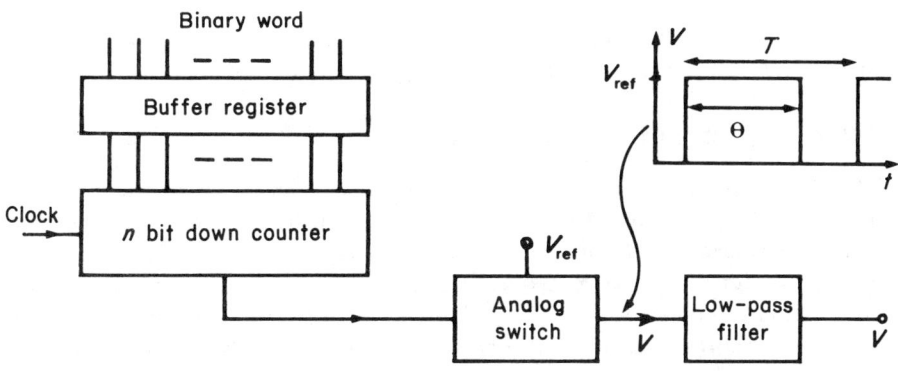

Figure 45

When a conversion is to be carried out, the binary word N to be decoded is transferred from the buffer register to the down-counter system. If none of the bits of this word are zero the clock pulses are applied to the down-counter and its content changes. As long as the content of the counter has not reached zero the analog switch is closed and the voltage V_{ref} appears at the output. As soon as the content reaches zero the analog switch opens and the input voltage to the low-pass filter is zero. Thus a pulse is obtained of amplitude V_{ref} and of width proportional to the number of pulses counted down, that is, to N. If T is the conversion period and θ the pulse width the continuous component of the output signal is

$$V = V_{ref} \frac{\theta}{T}$$

If T_H is the clock period then $\theta = N T_H$ and thus

$$V = V_{ref} N \frac{T_H}{T}$$

Usually the frequencies are adjusted so that $T = 2^n T_H$. The time needed for one conversion is therefore equal to $2^n T_H$. If for example the clock frequency is 1 MHz and $n = 10$ then about 1 ms is needed to carry out one conversion. Prior to filtering the signal is made up of pulses and its spectrum consists of an infinite number of lines spaced at intervals of $1/T$ whose amplitude decreases as a function of $(\sin x)/x$ where $x = n\pi\theta/T$; that is the rate of decrease is slower as θ is smaller. In order to retain only the continuous component, whose instantaneous value varies with N, it is necessary to eliminate the first line at frequency $1/T$. It is therefore necessary that the time constant of the output filter be much greater than the conversion period T, its value depending upon the accuracy required. If an accuracy of 10^{-3} is required the line at frequency F will have to be attenuated by a factor of 2000. If a double pole RC filter is used for that purpose then its cut off frequency $\omega_c = 1/RC$ must be such that $2\pi/\omega_c T = 33$. In the above example the system will be practically unable to accept signals of frequency higher than than 10 Hz. In order to improve the pass band a faster clock must be used (up to 500 MHz is realizable at present) or deal with fewer bits. Instead of producing a pulse with a width proportional to N, one may provide a train of pulses of constant width and amplitude but whose number is proportional to N. The arrangement used in this case is a binary multiplier instead of a down counter.

2.5.3.2 Stochastic converters

Another principle of operation of an indirect DAC makes use of the *stochastic representation* of the information to be converted. This representation, which has recently been examined for use with computers, has led to major circuit simplification for it is possible to discard nearly all the usual analog part used on DACs. It then becomes possible to integrate a whole DAC on one chip (LSI). In the stochastic representation the information is represented by the probability of a pulse occurring, and since this occurrence is random, its uncertainty leads to devices of great simplicity. The basic schematic diagram of a stochastic DAC[28] is given in Fig. 6. Both the n bit signal to be converted and the n bit signal produced by a random number generator are fed into a comparator. The comparator gives a 1 or a 0 depending upon the result of the comparison. If the generator is truly random all sequences have the same probability of occurrence.

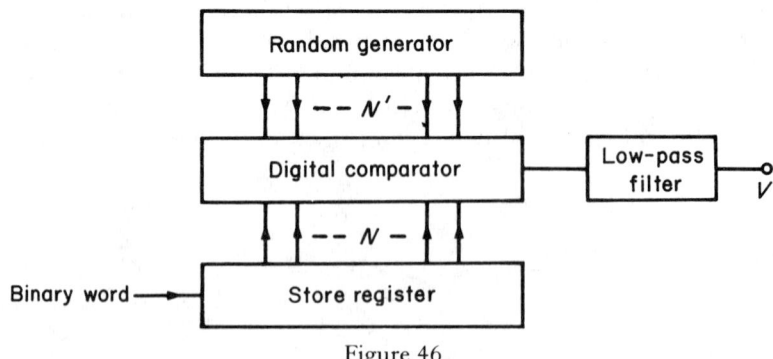

Figure 46

The principle of stochastic conversion consists, therefore, in the *measurement of the probability of occurrence* of certain combinations. In theory this measurement would require an infinite time so the accuracy of the measurement will depend on the length of time taken for the measurement and will improve as the length of time is made longer. Hence this type of converter is slow because of the very principle of its operation.

An example will clarify this operating principle. Consider a dice where the probability of occurrence for each side is $\frac{1}{6}$. If all the throws for which the number thrown is less than or equal to 2 are counted then the ratio of such throws to the total number of throws is $\frac{2}{6}$ provided the total number of throws is large. There exists, therefore, a simple linear relationship between the probability of obtaining one of the numbers on the dice and the number required. If the density is no longer uniform then the probability relation is no longer linear so that:

$$S = 1 \quad \text{if } B < N$$

S being the stochastic information, B being the signal from the random number generator and N the number to be converted and

$$S = 0 \quad \text{if } B > N$$

It follows that:

$$\text{Probability}(S = 1) = \text{Probability}(B < N) = \int_{-\infty}^{N} f(B) \, dB$$

$f(B)$ being the probability density of B. If linear encoding is to be achieved then it is necessary to have a *constant probability density* over the range of variation of the input variable N. The number of tries is restricted and only an estimate of the probability can be made, for which two methods may be used; either a counter may be incremented each time the result is positive or the average value of the stochastic signal taken using a low-pass filter.

This principle is usually applied as follows; the value N is compared to the finite number N' supplied by the generator and a pulse is sent each time N is greater than N'. If the conversion time is fairly long (in principle it should be infinite) the number of pulses sent (or the pulse density) is proportional to N. The average value is extracted by means of a low-pass filter whose output voltage is proportional to N. The difficulty resides in the manufacture of a noise generator which closely approximates a random generator. A pseudo-random generator can be fairly easily obtained by using flip-flops. For example one can use a binary counter in which the roles of the two end bits are interchanged, the MSB becoming the LSB and vice versa and this method lends itself conveniently to integrated circuit fabrication. The digital comparator may use a binary adder, taking into account the fact that the distribution law for N' does not change during the complementing operation. Thus

$$S = 1 \quad \text{if } N + N' > 2^n$$

Signal S will therefore be the carry bit from the adder. The whole advantage of this type of converter lies in the fact that integrated circuit fabrication can be used and hence profit gained from the vast experience in that field.

2.6   MEASUREMENTS OF THE CHARACTERISTICS OF A DAC

The methods and configurations used for testing a DAC[19,29,30] will depend on a number of factors such as the eventual use required of the converter, the nature and speed of the required tests etc. The relative importance of the specifications depends upon the applications and users may wish to check those parameters which most affect the overall performance of the system. For example, in the case of vector generators for cathode ray tubes the important parameters are the differential linearity, a fast conversion time and small switching transients. On the other hand if a DAC is used as a programmable generator in an automatic control system its operation demands good calibration and stable zero setting.

The automatic equipment capable of testing various types of converter is, of necessity, complex and costly. Luckily, there are simple and fast methods for measuring with sufficient accuracy the main performance criteria of a DAC, the *linearity* and *conversion time*, and which also include the effect of temperature on the linearity.

2.6.1   The Measurement of Linearity

This measurement is only meaningful under given external conditions of temperature, etc. The most accurate method of checking the *linearity* and at the same time measuring the *accuracy* of a DAC consists in checking all the possible combinations of the input binary code. (There are 2^n such combinations for an n bit word.) For a given combination the difference between the value of the voltage measured at the output and a theoretical value will give the accuracy of the converter. By plotting the transfer characteristic and comparing it to the ideal characteristic the linearity and monotonicity are obtained, but this method is time consuming.

The simplest but nevertheless a very effective method for checking the linearity requires only n stages for an n bit converter and can be carried out semi-automatically. It consists in *comparing the contribution of one bit with the sum of the contributions* of all the bits of lower weight using the identity:

$$2^i = 1 + \sum_{j=0}^{i-1} 2^j$$

The two results must therefore differ by a quantum. If the difference is positive the system is monotonic, if it is less than two quanta the differential linearity is less

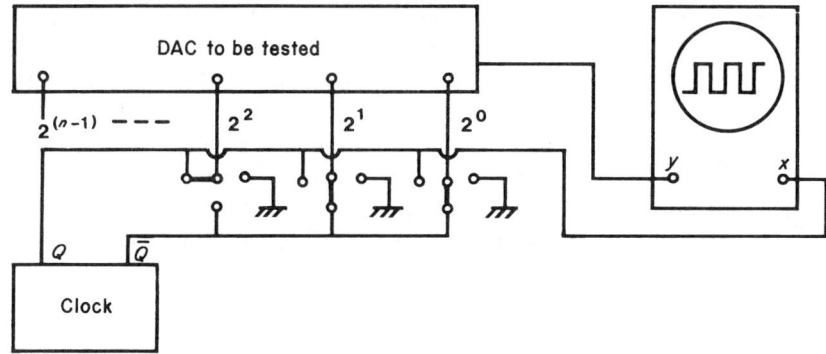

Figure 47

than $\pm\frac{1}{2}$ quantum. The circuit arrangement used is shown in Fig. 47. In this example the voltages corresponding to the words 100 and 011 are compared. The output signal of the converter is fed to an oscilloscope which is synchronized with a clock so that the difference between two successive tests is displayed on the screen. The system has n 3-way switches which control the state of each bit of the converter. The output signal Q or $\bar{Q}$ from the clock may be applied to the bits, or else they may be earthed and become inactive. The first $(i-1)$ bits are controlled by $\bar{Q}$, the ith bit by Q and those of rank in between $i+1$ and n are rendered inactive. Several possible cases are shown in Fig. 48. Signal A corresponds to 2^i and signal B to the sum of the lesser bits, $\sum_{j=0}^{i-1} 2^j$.

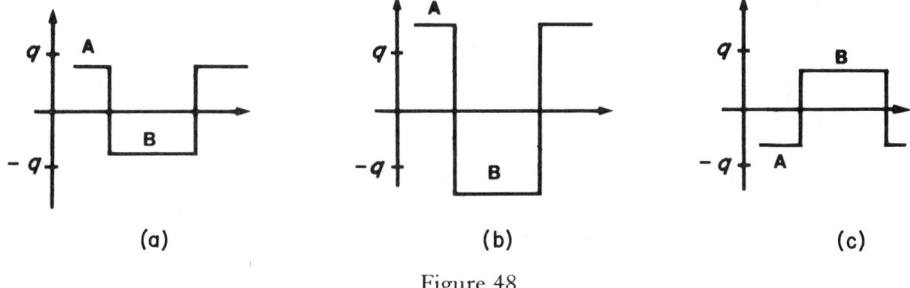

| (a) | (b) | (c) |

Figure 48

In the case of Fig. 48(a), the differential linearity error is less than $\pm\frac{1}{2}$ quantum and for Fig. 48(b) the error is greater than $\pm\frac{1}{2}$ quantum whereas Fig. 48(c) corresponds to a non-monotonic system (A $<$ B).

A second method of interest consists in using a *reference DAC* which must be more accurate than the DAC to be tested (at least three bits better). Both DACs are controlled by a binary counter of n bits and therefore supply the same signal to within the error of the converter under test. The signals are subtracted and the difference may be displayed on an oscilloscope. When the resolution of the converter is not too high (number of bits at most equal to eight) it is possible, by using a clock, to control the binary counter and also synchronize the oscilloscope to display the error of the converter under test. The oscillogram appears as a series of dots that must be aligned horizontally if there is no error. The deviation between the horizontal axis and the various points enables one to calculate the linearity error of the converter.

Other measurements can be carried out in order to establish the linearity and accuracy of a DAC. First, each bit is set individually to 1 (the others remaining at 0) and the voltage obtained for one bit must be equal to the sum of the voltages obtained for the bits of lesser weight to within one quantum. Alternatively one bit is set to 1 and all the other bits set first to the state 0 and then to the state 1. This enables one to examine the *independence* of the various bits (superposition error).

2.6.2 The Measurement of the Conversion Time

The conversion time is the time that the output signal takes to reach the value corresponding to the new input signal within a given accuracy (usually $\pm\frac{1}{2}$ quantum). There are two ways of specifying the conversion time which are of interest; the time corresponding to a full scale change, and that corresponding to the change of the LSB.

The measurement of the conversion time for full scale may be carried out using the circuit arrangement of Fig. 49. All the bits of the DAC are controlled by a

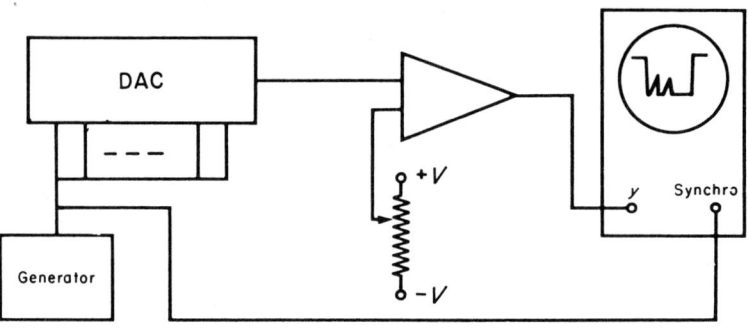

Figure 49

pulse generator which is also used to synchronize the oscilloscope. The output signal of the DAC is applied to one of the comparator inputs, while a reference voltage which can be varied over the whole dynamic range of the converter is applied to the other input. In this way the variation of the differential signal applied to the comparator is reduced and hence its response to a single step improved. Moreover the signals fed to the oscilloscope are not the output signals of the converter but reshaped signals for which the measurement of the conversion time is more conveniently carried out (Fig. 50). The high speed comparators that must then be used could degrade the information for they are temperature sensitive. (As a rule they have an input stage with high offset current in order to achieve a large gain-pass band product.)

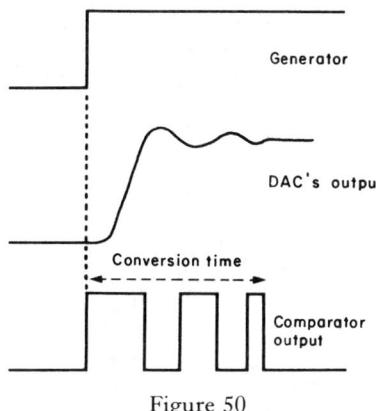

Figure 50

The conversion time corresponding to the change of one bit can be measured by means of the arrangement of Fig. 47. The point usually chosen is at the transition of 011 ... 11 to 100 ... 00 for which all the bits change state. The arrangement will also show if there are any transient switching conditions (glitches) or not.

2.7 EXAMPLES

As an example Table 5 (overleaf) reproduces the characteristics of two 12 bit DACs, one with voltage output and the other with current output. The aim of the examples chosen is to show above all how the manufacturers specify the characteristic of a DAC rather than recommend two particular converters.

It should simply be noted that the specifications are not always expressed in the same way (one speaks of conversion time or of speed of conversion) and sometimes the specifications are incomplete (the linearity is a case in point). When a user has to choose a DAC, they should first draw a specification of the requirements and then look for a DAC giving the best cost/performance ratio.

Table 5. Specifications (typical values at 25 °C if temperature is not stated)

	Manufacturer (type)	
	Analog Devices (DAC 80-CBI)	Teledyne Philbrick (4024)
Resolution (bits)	12	12

Input signals
  state 0
  state 1
	TTL compatible
	$E < 0.8$ V $-100\,\mu$A
	2VC $E < 5.5$ V $1\,\mu$A
		TTL compatible
		$E < 0.8$ V
		$E > 2$ V

Input code
  Unipolar
  Bipolar
	complementary binary,
	complementary, offset binary
	complementary 2s complement
		binary,
		offset binary

Output signals	Voltage: ± 2.5 V; ± 5 V; ± 10 V 0 to +5 V; 0 to +10 V Current: ± 1 mA; 0 to -2 mV	0 to +2 mA ± 1 mA
Output impedance	Voltage: 0.05 Ω Current: 6.6 kΩ	Unipolar: 8 kΩ, 40 pF Bipolar: 3.5 kΩ, 40 pF
Conversion speed (at 0.01% of full scale, ns)	Voltage: 5000 Current: 300	300
Rise time (V μs^{-1})	15 V	
Linearity	$\pm\frac{1}{2}$ LSB max. for 0–70 °C	$\pm\frac{1}{2}$ LSB max.
Differential linearity	$\pm\frac{3}{4}$ LSB max. for 0–70 °C	$\pm\frac{1}{2}$ LSB max.

Temperature coefficient
  (for 0–70 °C, ppm °C^{-1})
  gain
  offset unipolar
       bipolar
  differential linearity
	
	± 30
	± 3
	± 10
	
		± 30
		± 4
		± 18
		± 6

Supply	+15 V 10 mA -15 V 20 mA +5 V 8 mA	+15 V 15 mA -15 V 15 mA

Sensitivity of the supply
  (% of full scale per %
  change of supply)
	± 0.002	± 0.05

Standard temperature range (°C)	0–70	0–70

3 ANALOG TO DIGITAL CONVERSION

3.1 DEFINITION

An *analog to digital converter,* or ADC, is a device which receives an analog signal A, and converts it into a digital signal N with given accuracy and resolution, by comparing it to a reference voltage V_{ref}. In an ideal, error free ADC the output signal N is related to the input signal by the relation:

$$N \equiv \frac{A}{V_{\text{ref}}}$$

which amounts to saying that a division operation is carried out. This is usually done using decreasing powers of two in order to express the result directly as a binary number. In fact, since the transmitted message has a finite length, N can be considered to be the closest approximation to the result taking into account the resolution of the system.

The signal A to be converted can be written as:

$$A = V_{\text{ref}} \left(\frac{b_1}{2} + \frac{b_2}{4} + \ldots + \frac{b_n}{2^n} + \frac{b_{n+1}}{2^{n+1}} + \ldots \right)$$

since the length of the binary word obtained after conversion is limited to n bits, the division of A by V_{ref} must stop at the nth term and thus

$$A \simeq V_{\text{ref}} \left(\frac{b_1}{2} + \frac{b_2}{4} + \ldots + \frac{b_n}{2^n} \right)$$

the neglected terms b_{n+1}, b_{n+2}, $\ldots$ represent the conversion error, also called the quantization error. A–D conversion is therefore a *quantization operation.* This operation consists in replacing the voltage A by a discrete voltage which is a multiple of an elementary quantity called a quantum of value $V_{\text{ref}}/2^n$ so that the modulus of the difference between A and that discrete voltage is less than half a quantum:

$$\left| A - N \frac{V_{\text{ref}}}{2^n} \right| \leq \frac{1}{2} \frac{V_{\text{ref}}}{2^n}$$

If the maximum amplitude of the input voltage to be encoded is represented by a vector, then the quantization operation divides it into a finite number of segments all one quantum in length (that is $V_{ref}/2^n$), each being assigned a number in the chosen code (usually the binary code). Thereafter a segment is identified by its number and in order to convert a voltage A it will only be necessary to indicate the number of the segment in which the tip of the vector representing its amplitude lies.

The very nature of quantization makes it difficult to establish explicit relations between the input and output signals of an ADC. However the *transfer function of an ADC* can be defined as follows:

the nominal input voltages (those for which the error is zero) are given by:

$$E_{nom} = V_{ref}\left(\frac{b_1}{2} + \frac{b_2}{4} + \ldots + \frac{b_n}{2^n}\right)$$

and the various b_i are 0 or 1.

the range of input voltages which gives a specific value of the output signal satisfies the inequalities:

$$E_{nom} - \frac{1}{2}\frac{V_{ref}}{2^n} < A < E_{nom} + \frac{1}{2}\frac{V_{ref}}{2^n}$$

It has been implicitly assumed throughout the above discussion that the voltage to be converted was positive. However there are instances where the voltage can be positive or negative and a bipolar code must then be used (magnitude-sign or offset binary code). It will be shown when the various types of ADCs are examined that they can all accept bipolar voltages after suitable modifications.

3.2 CLASSIFICATION OF ADCs

The classification of DACs raised no special problems since the various methods used followed directly from the definition of the transfer function, but this is not the case for ADCs[31] which provide a much more varied choice and consequently the question of classification is much more subtle.

First, they could be classified into direct or indirect ADCs. In direct ADCs the digital signal is established directly by comparing, for example the analog signal with a series of weighted voltages. In indirect ADCs the input signal is transformed into an intermediate signal (a time interval for instance) which is itself transformed into a digital signal.

A second criterion distinguishes between ADCs with or without feedback. Feedback type ADCs more often than not include a DAC in their feedback loop. In this case the digital output signal is converted into an analog signal by a DAC and fed back to the input to provide an error signal if the conversion is not satisfactory. Different bits are thus determined at different instants of time.

Converters with feedback loop can themselves be subdivided into two categories: sequential or serial converters in which one bit is converted per period and serial–parallel converters in which two or more bits are converted per period. Another possible classification criterion available for indirect converters is based on the quantization signal. The intermediate signal may take the shape of a pulse whose width is proportional to the signal to be converted or it may be a periodic signal whose frequency is proportional to the amplitude of the analog signal.

These classifications are not mutually exclusive and it may happen that an ADC may well be classed under several category headings. Clearly then it is difficult to choose what seems the best classification from the possibilities mentioned above.

The classification finally adopted is the following: converters are divided into *analog* type and *logic type ADCs*. The first family covers all ADCs whose operation is basically analog, that is to say they use analog techniques to arrive at a solution such as the generation of a ramp voltage, the charge of a capacitor etc., while on the other hand the second family covers converters whose operation depends largely on digital techniques. *Logarithmic ADCs* which can belong to both families will be treated separately as will *fast* and *very fast* ADCs which require individual and more complex arrangements.

3.3 CHARACTERISTIC PARAMETERS OF AN ADC

The main characteristics of an ADC[2,17–20] bear the same names as those of a DAC but they differ in the way they are defined since the type of input and output signals are interchanged. Unless otherwise stated they are defined for a unipolar code.

3.3.1 Ideal Transfer Function of an ADC

The ideal transfer function of an ADC is defined by the following two expressions:

$$E_{nom} = V_{ref}\left(\frac{b_1}{2} + \frac{b_2}{4} + \ldots + \frac{b_n}{2^n}\right)$$

$$E_{nom} - \frac{1}{2}\frac{V_{ref}}{2^n} < A < E_{nom} + \frac{1}{2}\frac{V_{ref}}{2^n}$$

V_{ref} represents the magnitude of the voltage to be converted, also called full-scale range. The transfer characteristic of an ADC is shown in Fig. 51(a) and it can be seen that:

it consists of steps of width $V_{ref}/2^n$ whose mid-points correspond to the various values of the voltage E_{nom},

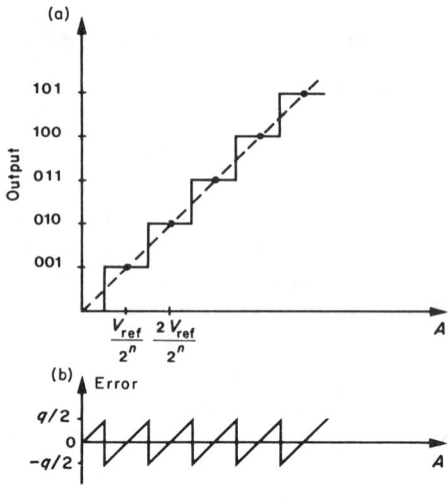

Figure 51

these various mid-points lie on a straight line which is called the ideal characteristic of an ADC,

the transition steps occur for voltages $E_{nom} + \frac{1}{2}(V_{ref}/2^n)$. Figure 51(b) shows the corresponding conversion error which represents the difference $A - E_{nom}$, assuming a constant quantization. The modulus of this error is smaller or equal to $\frac{1}{2}(V_{ref}/2^n)$ and the limit of this error is known as the *quantization error*. The conversion error is zero for the values E_{nom}.

Since N is $2^n - 1$, the difference between the value of the voltage that can be converted and the voltage V_{ref} is one quantum.

3.3.2 Resolution

Its definition is the same as that given for a DAC. It is fixed by the number of bits given by the converter. The resolution specifies the smallest voltage change that can be encoded and detected by the converter taking into account the number of bits. The resolution is usually measured in terms of full scale, and when expressed as a relative quantity it is given by:

$$r = 1/2^n$$

3.3.3 Conversion Time

The conversion time is the time required to obtain a digital output signal corresponding to the analog input signal within a specified accuracy. Usually it is expressed in microseconds or in milliseconds. This definition assumes that the converter is *ready to operate* when the timing of the conversion is started. To obtain the *maximum conversion frequency* possible the time taken to reset the converter to zero must be added to the conversion time. Usually this time is fairly small and does not reduce the performance of the system appreciably, except for the ultra fast converter. It is also assumed that the value of the input signal has not changed during the conversion time, otherwise no significance could be attached to the measurement.

The conversion time is usually specified for the maximum variation of input voltage. In certain systems, in particular for sequential ADCs, this time is independent of the value of the signal and only depends upon the number of bits and is therefore always the same. On the other hand, for other systems this time depends upon the value of the signal to be converted, and if the user knows that the amplitude of the signal will never reach the maximum permissible value they can calculate the maximum conversion time for the application and thus increase the conversion rate.

Sometimes the manufacturer specifies the maximum conversion frequency which is not exactly equal to the inverse of the conversion time. It corresponds to the maximum number of conversions that can effectively be carried out in one second and takes into account various dead times, notably the time required to reset to zero.

3.3.4 Accuracy

The accuracy of an ADC is defined as the difference between the theoretical value of E_{nom} producing a given word N at the output and the real value of A required to produce this word. This difference is called the absolute error and can also be expressed as a percentage or as a fraction of a quantum when it is referred to as the relative accuracy or relative error. The main sources of errors are quantization, the zero offset, change in gain or scale factor and non-linearity.

3.3.5 Noise Rejection

It is a feature of some types of ADC that they reduce or even eliminate the effect of some types of noise, notably mains supply noise. In this case a voltage noise rejection factor $S(\omega)$ is defined as the ratio of the normalized input noise $A_{(noise)}/V_{ref}$ to the normalized output noise $N_{(noise)}$

$$S(\omega) = \frac{A_{(noise)}/V_{ref}}{N_{(noise)}}$$

This factor is frequency dependent and is of particular importance for low speed ADCs, for in this case the mains supply noise can introduce important errors (notably if the ADC includes integrating systems; for fast ADCs this factor is not so important). It will be shown that certain principles of A–D conversion can reduce considerably the effects of parasitic noise provided certain precautions are taken.

3.3.6   Bipolar ADCs

The preceding definitions must be slightly modified when dealing with a bipolar converter.

(1) The expression for the ideal transfer function depends upon the code used. For example, for the binary code it is:

$$E_{nom} = 2V_{ref}\left(\frac{b_1}{2} + \frac{b_2}{4} + \ldots + \frac{b_n}{2^n}\right) - V_{ref}$$

On the other hand the second expression defining the transfer function remains the same as for unipolar ADCs, namely:

$$E_{nom} - \frac{1}{2}\frac{V_{ref}}{2^n} < A < E_{nom} + \frac{1}{2}\frac{V_{ref}}{2^n}$$

This expression stating the fact that the conversion is carried out with a quantum equal to $V_{ref}/2^n$.

(2) The shape of the transfer characteristic is altered, becoming symmetrical with respect to the origin, ranging from $-V_{ref}$ to $+V_{ref}$.
(3) The variation of the conversion error as a function of the signal to be converted remains the same.
(4) Full scale, that is the range of the voltage amplitude to be converted, is approximately doubled and is written as:

$$A_{max} - A_{min} = 2V_{ref} - \frac{V_{ref}}{2^n}$$

(5) As was the case for DACs, if the full scale is to be doubled while keeping the same quantum value (in absolute terms) then a bit must be added.

3.4   ERRORS IN ADCs

Generally the actual performance of an ADC differs from the performance predicted by theory. Instead of having the shape of the ideal characteristic of Fig. 51(a) the actual transfer characteristic can have the shape shown in Fig. 52 and the following differences should be noted.

The voltages at which transitions occur differ from the corresponding voltages on the ideal graph.
The width of the horizontal steps on the actual graph is not equal to the theoretical value of one quantum.
The error does not remain within the interval $-q/2$, $+q/2$ (Fig. 52(b)).

These differences are essentially due to: the *gain error*, the *offset error* and the *linearity error*, to which must be added the quantization error. Each error[19,20,32]

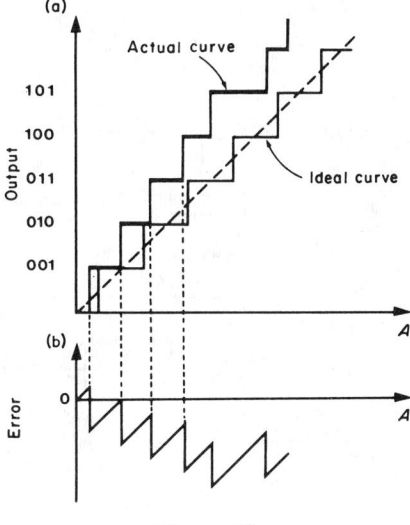

Figure 52

will be defined separately under the assumption that the other errors have been eliminated.

3.4.1 Quantization Error

The first error which is met in an ADC is the *quantization error*. This error does not occur in DACs but is *inherent* to the principle of analog to digital conversion (which is in essence non linear) and whatever the method of conversion used always has the absolute value of a half quantum. Its value depends only upon the resolution of the system. In contrast to the errors which are defined next, the quantization error can be considered as a theoretical error, while the others are due to manufacturing imperfections.

3.4.2 Offset Error

The *offset error* is the difference between the value of voltage applied at the input which sets the LSB into state 1 and the theoretical value of the voltage which will bring about this change (given by $\frac{1}{2}(V_{ref}/2^n)$). This voltage difference is called the *offset voltage*. It results in a horizontal translation of the graph of the transfer function and a vertical displacement of the conversion error which is then symmetrical about the offset voltage (Fig. 53). It is possible to eliminate the effect of the offset voltage for a given temperature. Usually this is achieved by introducing into the circuit an opposing offset voltage by means of an operational amplifier. Unfortunately the parameters which are the source of this offset voltage are temperature dependent so that the error will reappear if the temperature

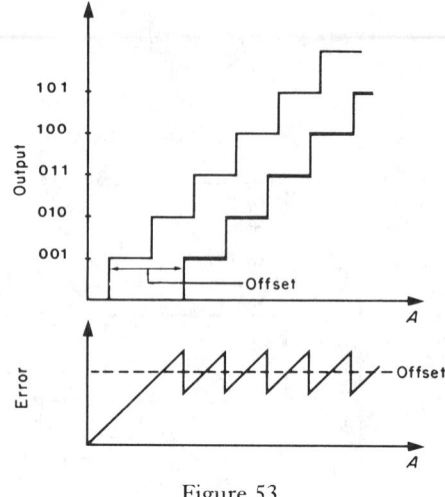

Figure 53

changes. It is generally very difficult to compensate this error over a wide temperature range.

3.4.3   Gain Error

The *gain error* (also called the scale factor error) results in the line joining the points representing the voltages E_{nom} rotating about the origin (Fig. 54). One may then write:

$$E_{\text{nom}} = k\,V_{\text{ref}}\left(\frac{b_1}{2} + \frac{b_2}{4} + \ldots + \frac{b_n}{2^n}\right)$$

k being a positive coefficient greater or smaller than one. If $k > 1$ the steps become smaller and the error increases with the input voltage but 'saturation' occurs before the value V_{ref} is reached. If $k < 1$ the steps are larger, the error also increases and it is not possible to obtain the output number $11 \ldots 11$, corresponding to the maximum voltage.

This error can be eliminated at a given temperature by adjusting the overall gain of the converter to unity. As was the case for the offset error it will reappear if the temperature changes.

3.4.4   Linearity Error

As was the case for the DACs the *linearity error* is difficult to define and is open to several interpretations. If the transfer characteristic was a continuous curve the linearity error could be defined as the difference between the actual curve and the theoretical curve for a given input voltage. In fact this curve is only defined at the values of A resulting in a transition step and for values of E_{nom} which are located

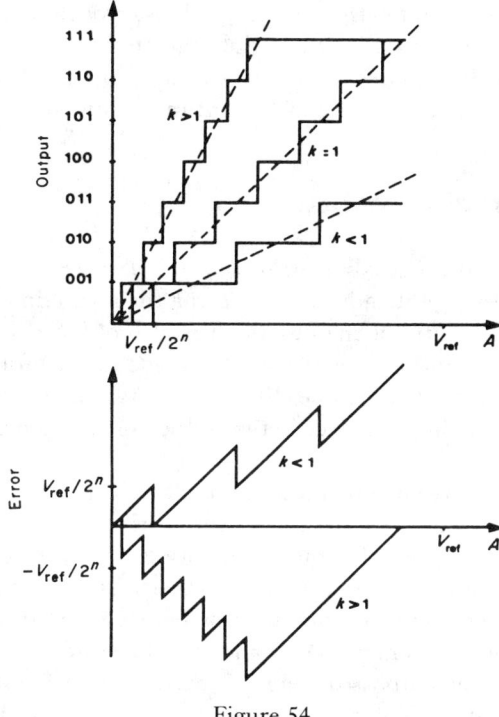

Figure 54

theoretically at the mid point of the horizontal segments. Under the assumption that there are no gain and offset errors, the linearity gain is defined as the difference between the values of voltages A, which effectively ensure the transition steps and the corresponding theoretical values, that is $V_{ref}/2^n(N+\frac{1}{2})$ (Fig. 55). It can also be said to be the difference between the values of E_{nom} and the voltages of the mid points of the various horizontal segments.

Some authors define the linearity error, in the presence of a gain error, as the difference between the actual curve and the 'best fit' straight line joining the extreme ends of the actual curve. Best fit is in quotation marks because this definition does not seem very realistic in the sense that the user is not interested in

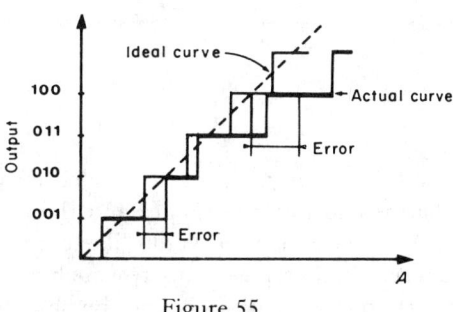

Figure 55

part of the linearity error but in the total error. This error cannot be compensated for since it results from the variations in the characteristics of all the component parts of the converter with the amplitude of the input signal. It is usually given as a fraction of a quantum and a value of $\frac{1}{2}$ quantum is quite usual.

3.4.5 Differential Linearity

The linearity discussed above is sometimes called *integral linearity* for even if it is expressed as a percentage it indicates the error for a certain expression of the output word independently of the others. It is also of interest to compare two consecutive states and in this case reference is made to the differential linearity. The *differential linearity* error is the difference between the quantum and the width of a step (that is the difference between the voltages causing two consecutive transition steps).

For example, in the case of a quantum of one volt, if the transition step occurs at 2.3 V and the next step at 2.9 V the differential linearity error is: $1 - (2.9 - 2.3) = 0.4$ V. If the differential linearity error is greater than one quantum, then certain combinations can never appear at the output. This is the inverse phenomenon of the non-monotonicity defined for DACs. This may occur if the ADC uses a non-monotonic DAC and it is then said that certain combinations are missing. Some authors distinguish between non-monotonicity and the fact that certain combinations cannot be obtained. When certain combinations are missing but the transfer characteristic may be replaced by a straight line segment of positive slope the system is called monotonic and non-linear (Fig. 56(a)). On the other hand when the output number decreases while the

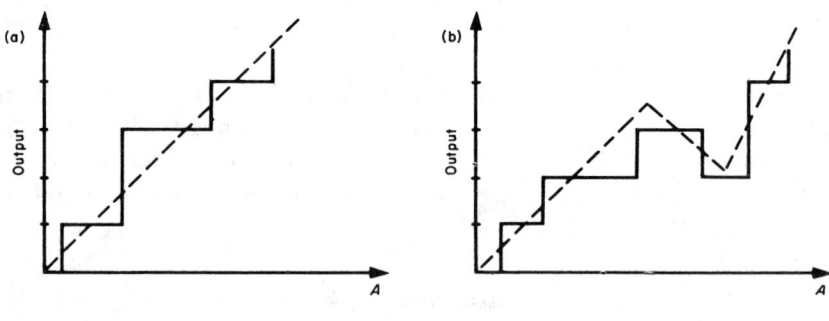

Figure 56

input signal increases the characteristic may be replaced by a number of straight line segments of positive and negative slopes and the converter is called non-monotonic (Fig. 56(b)). In fact the user would like to know the overall accuracy of the converter and it seems, therefore, more reasonable to say that the converter is not monotonic if certain combinations cannot be obtained at the output.

3.4.6 Temperature Effects

As was the case for DACs, certain errors defined above vary with temperature. These are the gain, offset and linearity errors. The changes in these errors are added to the already existing known errors at ambient temperature and thus the performance of a converter can be predicted over a specified range of temperatures. They are usually specified in $\% \,^{\circ}\text{C}^{-1}$ and more seldom in $\mu\text{V}\,^{\circ}\text{C}^{-1}$. The additional error due to changes of temperature is calculated as for the DACs and one must not forget to add to all these errors the effect of supply voltage variations which is sometimes temperature dependent. However in ADCs temperature is sometimes the source of other errors. Firstly the power dissipation of certain types of integrated logic circuits cannot be ignored and results in an increase of the working temperature of all or part of the converter, thus modifying its performance. Moreover certain ADCs use an internal clock made by simply looping together a few gates, but the capacitances, saturation currents, etc. and hence the clock frequency vary with temperature. In this case a change in the conversion time occurs which can have serious consequences if the time allowed for each operation is arrived at through a constant internal delay.

3.5 ANALOG ADCs

Under this heading are grouped all the ADCs for which the operating function is largely analog, for example the transfer of charge from one capacitor to another, the generation of a linearly varying voltage etc. The most important and better known types will be examined, paying particular attention to the different sources of possible errors.

3.5.1 Single-slope Converters

This is one of the simplest and most popular converters[33,34] for it requires few components. It is also called a pulse width modulation converter, because a pulse is produced whose width is proportional to the value of the input voltage. The schematic diagram for such a converter is shown in Fig. 57(a) and the corresponding timing diagram on Fig. 57(b).

Before the start of the conversion operation, switch S is closed, discharging capacitor C. At time t_0, the start of the conversion operation, the start command signal resets an n bit counter to zero and opens switch S. A constant current source I then charges capacitor C and the voltage ramp which is obtained is applied to one of the inputs of a comparator whose other input is connected to the voltage V_x to be converted. (To a first approximation, a comparator can be considered as an amplifier of infinite gain and whose output voltage can only have two values and which changes state when the difference between its input voltages

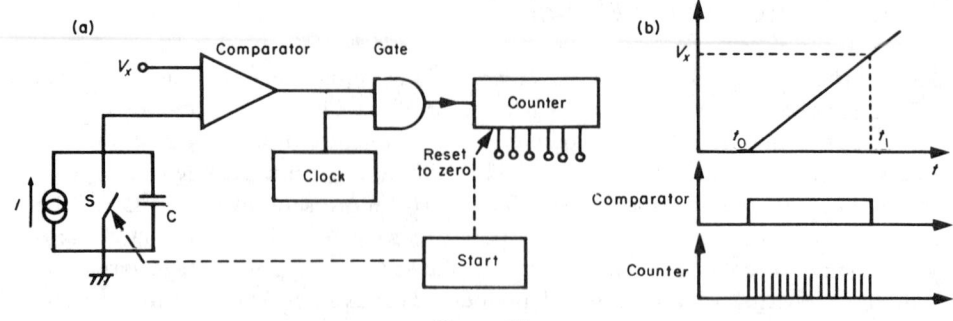

Figure 57

changes sign.) Simultaneously the gate which is enabled at the start of the conversion operation, feeds the clock pulses of period T to a counter. When, at time t_1, the capacitor voltage reaches the value V_x the comparator changes state, the gate closes and the counter stops counting since it receives no further pulses. The ramp voltage is then

$$V_c = \frac{I(t_1 - t_0)}{C} = V_x$$

and the counter has recorded a number of pulses $N(0 \leqslant N \leqslant 2^n)$, so that

$$V_x = \frac{1}{C} NT$$

The number of pulses recorded is proportional to the voltage V_x to be converted and can therefore be seen as a form of encoding of voltage V_x. As this number is held in the counter it is immediately available in binary form.

Since encoding V_x amounts to measuring a period of time (in fact the interval $t_1 - t_0$ is measured) this converter is also called a voltage–time converter. Conversion is achieved through the intermediary of the measurement of an interval of time and so this is regarded as an indirect converter. Because N can be equal to $2^n - 1$ it is necessary to be able to count $2^n - 1$ pulses, and this will take a time $t = (2^n - 1)T$. This value of t is the conversion time of the system. Thus for a clock frequency of 1 MHz ($T = 1 \, \mu s$) and a 10 bit counter (giving an accuracy of 0.1%) the system takes about 1 ms to convert the maximum permissible voltage. It is therefore rather a *slow* system.

The quantization error (theoretically less than $\frac{1}{2}$ quantum) can in this case reach 1.5 quanta. This happens if the start signal occurs just before the rise of the clock pulse which triggers the counter and immediately the counter will have a 1. This error will be added on to the usual quantization error. To overcome this drawback the start of the capacitor charging period can be synchronized with the clock pulses. These pulses can themselves be delayed by half a period before being directed to the counter. The transition steps are then produced at instants of time corresponding to odd multiples of $q/2$ (q being the quantum) in accordance with theory.

The system still has a rather average accuracy level (10^{-2} to 10^{-3}) and the main sources of error are easily deduced from the expression of V_x in terms of N. One of the most important factors affecting the accuracy is the linearity of the sawtooth generator (that is the constancy of current I). If the sawtooth part is not linear then an error will occur whose value will change with V_x. This error can be reduced if an integrator comprising an operational amplifier is used, often at the cost of increasing the conversion time. Equally the stability of the clock period affects the overall accuracy and attempts must be made to reduce its short and long term drift effects. An additional factor is the accuracy to which the comparison is made. A parasitic voltage, superimposed on voltage V_x could significantly upset the result by changing the time at which comparison takes place. The capacitance of capacitor C must not change with time otherwise an additional error will have been introduced. Finally the actual starting time of the sawtooth is uncertain as the initial part of the wave shape is nonlinear, and this reflects on the accuracy of the evaluation of t. This last error can be eliminated by using two comparators and by making the ADC suitable for bipolar signals (Fig. 58). The ramp and the unknown voltage V_x are applied to the inputs of

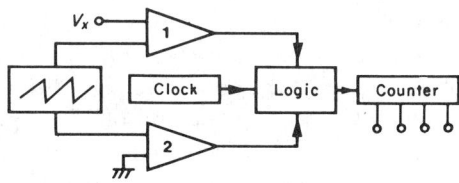

Figure 58

comparator 1. The same ramp voltage and a zero voltage are applied to the inputs of comparator 2. In this case the ramp voltage must vary from $-V_{ref}$ to $+V_{ref}$. The operation is the same as before, the output state of comparator 1 changes as soon as the ramp voltage reaches the value V_x and comparator 2 acts likewise when the ramp passes through the value 0. The first one whose output changes state enables the logic gate, and the second one to change inhibits it. During the enabling time the clock pulses are fed to the counter, which counts a number of pulses proportional to the absolute value of voltage V_x. A simple logic circuit establishes the sign of voltage V_x once it has detected which comparator has changed its output state first. The attraction of the ramp detector is essentially one of economics (it is rather slow and inaccurate) since the use of integrated circuits makes it a *low cost device*.

3.5.2 Dual Slope Converters

It is possible to increase appreciably the accuracy of the previous system by carrying out *a double integration*. Moreover, an even better rejection of parasitic signals is obtained, in particular of those originating from the mains supply.

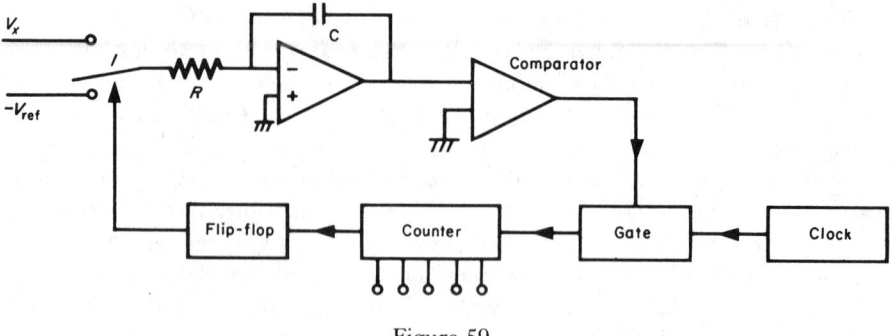

Figure 59

Figure 59 shows the schematic diagram for this converter.[35–37] This time the analog section has an integrator followed by a comparator, and the logic section is increased by one flip-flop. In order to explain the operation of this converter it is assumed that the signal V_x to be converted is positive but the principle is also applicable to bipolar signals. At the start of the conversion operation capacitor C is uncharged, the counter is set to zero and voltage V_x is applied to the integrator. The gate is enabled and the clock pulses are fed to the counter. Provided this has not reached its maximum capacity, the output voltage of the integrator decreases according to the relation $-V_x t/RC$. At the end of $2^n - 1$ pulses the maximum capacity of the counter is reached. The following pulse resets it to zero and sets the additional flip-flop in state 1. The flip-flop then operates the change-over switch I, which applies the voltage $-V_{ref}$ to the integrator input, which in turn supplies a rising voltage according to $V_{ref}t/RC$. The comparator detects the passage of this voltage through zero and inhibits the gate, stopping the pulses getting through.

During the period of time of the second integration the counter operates normally and counts N pulses of period T. This operation is described schematically in Fig. 60 showing the graphs of the variations of the output voltage of the integrator against time for two voltage values $V_{x1} < V_{x2}$.

It will be noticed from this graph that the time of the first integration is constant and corresponds to 2^n pulses and that the slope of the second integration is also

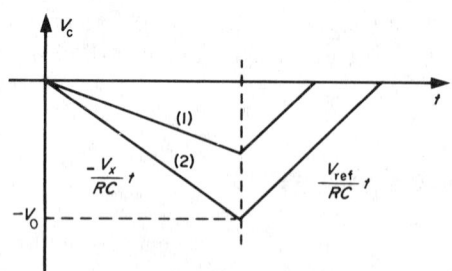

Figure 60

constant. Hence:

$$V_0 = \frac{V_x}{RC} 2^n T \quad \text{and} \quad V_0 = \frac{V_{ref}}{RC} NT$$

that is

$$V_x = \frac{V_{ref}}{2^n} N$$

The only variables appearing in the result are the reference voltage and the number of pulses counted, N. Consequently the accuracy of the system can be very high since possible capacitor or clock frequency drifts do not appear explicitly in the result, and affect both integrations in the same way. The differential linearity can be extremely good since the analog process has no discontinuity. This is explained by the fact that all the combinations are generated by the clock and the counter and, in principle, can all exist. The conversion time for this convertor is long since it must be able to count $2(2^n - 1)$ pulses, that is twice the number for a single-slope converter.

Finally the integration greatly reduces the effect of any noise that may be present in the input signal, whereas for a single-slope converter the noise which is superimposed upon the input signal can alter the comparison time and hence degrade the result because of the errors introduced.

If the rejection factor $S(\omega)$ of the disturbing signal is calculated in terms of frequency, the following relation is obtained:

$$S(\omega) = \sin\left(\frac{\omega\theta}{2}\right) \bigg/ \left(\frac{\omega\theta}{2}\right) \quad \text{where } \theta = 2^n T.$$

The graph of this equation plotted on a log–log scale is shown in Fig. 61. The envelope is a straight line of 20 dB per decade slope. $S(\omega)$ is zero for whole multiples of $1/\theta$ and hence rejection is perfect (or 'infinite') at these values.

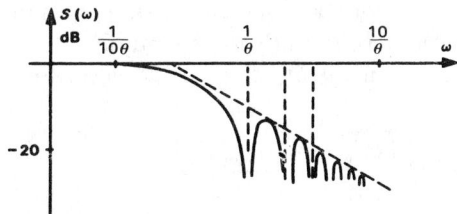

Figure 61

Usually this leads to very low conversion frequencies. For example if the effect of the 50 Hz supply frequency is to be eliminated then $\theta = 20$ ms and in this case the maximum number of conversions per second is 25. The great attraction of this system is its accuracy, which in the best cases can reach 10^{-5}. It is often used in digital voltmeters on account of its good noise rejection.

If the accuracy of the system needs to be increased then the conversion time must be increased in such proportions as to become prohibitive. It is therefore advantageous to carry out the conversion in two stages by having two measurements 'coarse' and 'fine'. In a first stage the n_1 MSBs are determined and the result is reconverted into an analog signal by means of a DAC. This signal is subtracted from V_x and the correctly amplified result is in turn converted into n_2 bits (where $n_1 + n_2 = n'$). Compared to an n bit converter, this converter has an accuracy $2^{n'-n}$ times better ($n' - n$ bits more) but the conversion requires a longer time. This system is sometimes called a '2-sectioned' converter. Usually n_1 and n_2 are made equal, so as not to have partial conversion times that are too different which would otherwise destroy the advantage of the system. It is possible to think of other methods of vernier measurements in particular by combining two different conversion techniques.

3.5.3 Counter Ramp Converters

The principle of this converter brings to mind that of the single-slope converter since a voltage is generated which increases in steps (having the shape of a staircase) and hence can be likened to a ramp. An advantage of this system is that time is no longer taken as a variable. In fact the variable of interest is the value reached by this changing voltage during a clock period, but not the exact instant of time of when it is reached. The schematic diagram for this converter is shown in Fig. 62(a). The analog signal V_x which is to be converted is compared with the output voltage of a DAC whose input signal consists of the n output signals available from a binary counter of 2^n capacity. At the start of the conversion process the clock pulses are counted by the counter and therefore it delivers all the possible combinations of n bits in increasing order. The DAC which receives these signals supplies a voltage V_0 which increases by one quantum each time the counter count increases by one unit. As long as this voltage is less than V_x the counter will continue to count and the output voltage of the DAC increases. When it exceeds the value V_x the comparator changes state, thus inhibiting the gate and this stops the pulse count (Fig. 62(b)). The number held in the counter corresponds to the binary equivalent of V_x since according to the principle of operation:

$$\left| V_x - \frac{N V_{\text{ref}}}{2^n} \right| \leq \frac{1}{2} \frac{V_{\text{ref}}}{2^n}$$

The attraction of such a converter lies in the fact that only two components affect the accuracy namely the DAC and the comparator. In practice the accuracy of such a system is determined by the DAC. The conversion time of this ADC is fairly long, just as is the conversion time of a single-slope converter, for according to the principle of operation 2^n pulses must be counted for each conversion. Moreover it is necessary to have a waiting period at each pulse before generating the next pulse while the voltage supplied by the DAC stabilizes (end of transient conditions) and the comparison between V_x and the DAC output voltage

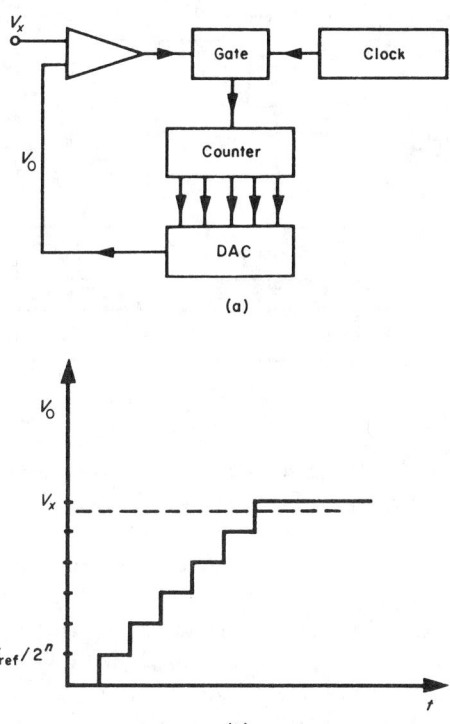

(a)

(b)

Figure 62

is made. Therefore the pulse period is at least equal to the sum of the DAC's settling time and the comparator's response time. The exact value of this period has no effect upon the result of the conversion operation and a clock with a less accurate period than is needed for a single-slope converter can be used (an accuracy only 1% for example). When starting a new conversion operation it is only necessary to reset the counter to zero to trigger the operation, since the DAC output signal automatically follows the changes in the counter contents. If the accuracy of this converter is to be increased by increasing the number of bits then the conversion time increases very rapidly. As in the case of the dual slope converter it is possible to avoid this drawback by carrying out the conversion in two stages by using two counters with a total capacity of n' bits (Fig. 63). At the start of the conversion operation only the first counter of n_1 bits corresponding to the MSB counts up the clock pulses. The output signal of the DAC has therefore a staircase shape each step being equal to $V_{\mathrm{ref}}/2^{n_1}$. When the comparator associated with the first counter changes state, the counter stops counting and the conversion of the first n_1 bits is completed. The clock pulses then are directed to the second n_2 bits counter (with $n' = n_1 + n_2$) corresponding to the LSB.

The voltage output of the DAC again increases in steps but the amplitude of the step is in this case equal to the value of the quantum needed to obtain the required accuracy, that is $V_{\mathrm{ref}}/2^{n'}$.

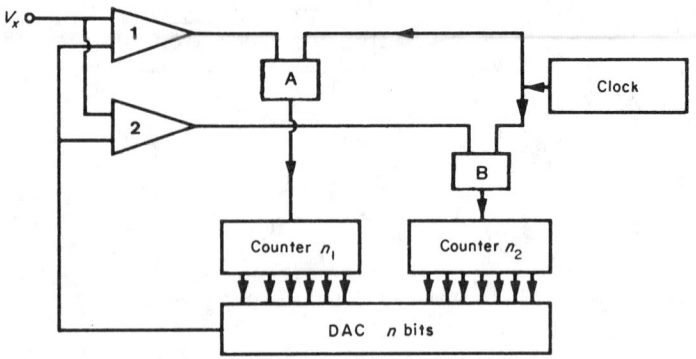

Figure 63

Compared with the previous converter this system requires additional logic circuits shown schematically on Fig. 63 as blocks A and B in order to control the alternating operation of the counters and to reset them to zero simultaneously. The conversion operation stops when the second comparator changes state, thus indicating that the output voltage of the DAC has reached the value of the unknown voltage V_x to within half a quantum. There is an optimum sharing of the bits between the two counters, for if n_1 is small the first steps are large and many small steps equal to the quantum $V_{ref}/2^{n'}$ (hence many clock periods) will be required to obtain the result. On the other hand if n_1 is large the conversion for the first part requires a time that could be too long.

3.5.4 Continuous Counter Ramp Converters

A limitation of the counter ramp converter, whether using one or two counters, is due to the fact that the counter is reset to zero at the start of each conversion operation. Even if the difference between the signals of two consecutive conversions is small provision must still be made for counting 2^n pulses. The *continuous counter ramp converter* (Fig. 64) does away with this drawback by using an up–down counter, and it takes its name from the fact that it uses a feed back loop whose function is to annul the error signal continuously. The error signal handled by the loop is the difference between the voltage V_x to be converted and the output voltage V_0 of a DAC. Depending on whether this signal is positive or negative the comparator output will be a logic signal of 0 or 1 and the clock pulses through circuits 1 and 2 will go to the up or down counter and increment or decrement it respectively. Thus the output signal of the DAC always follows the changes of the input signal by means of the content of the counter. If the changes in the input signal are small the time taken to produce the new binary representation of the input signal is considerably reduced. However if the input signal varies greatly, or during the first conversion operation, the system needs to be able to count 2^n pulses and for this reason its conversion time remains the same as that of a counter ramp converter.

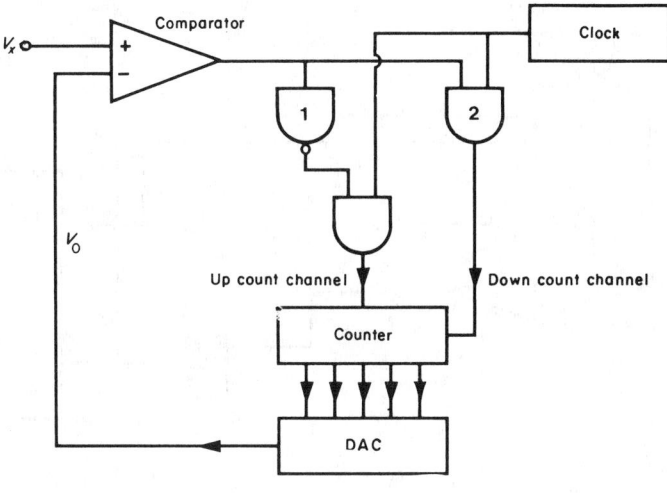

Figure 64

If the signal is constant, or if its variations do not exceed the value correspond-ing to the accuracy of the system ($\pm\frac{1}{2}$ quantum), the counter's content will tend to oscillate between two consecutive combinations. This drawback can be elim-inated at the price of a slight increase in the logic circuitry of the system. As long as the conversion operation is not completed the content of the counter may alter. A buffer memory can be placed after the counter which only provides this information when the conversion operation is completed, either through a signal originating from the logic circuitry or through some external agency. This simple and relatively inexpensive system is advantageous when one converter is to be used for each channel that is to be encoded, as in the case of multiplexing, and when the frequency of the signals is relatively low (for instance less than 100 kHz). Its accuracy largely depends on the accuracy of the DAC used. The system can also be used as a peak or trough indicator, or for sensing the direction of the changes of the analog signal.

3.5.5   Voltage to Frequency Converters

This converter[2,38] is not strictly an ADC, according to the definition given at the start of this chapter, but its operation closely follows that of a counter ramp converter. Moreover by simply adding a counter to it the whole arrangement does then constitute an ADC.

A voltage to frequency (V–F) converter can be defined as follows. It is a controlled oscillator whose output signal has a frequency proportional to the input voltage. If the frequency of the signal is measured by a counter then a digital indication of the input signal can be obtained.

The development of this converter was mainly due to the possibilities opened up by integrated circuits. A schematic diagram is shown in Fig. 65(a) and uses

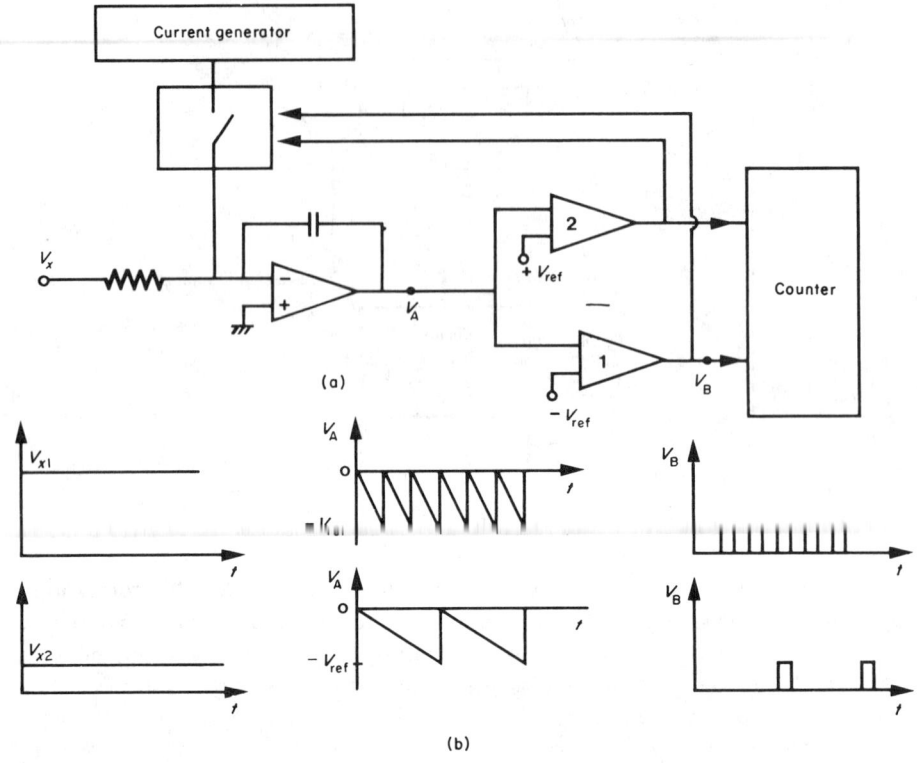

Figure 65

principally one integrator, two comparators and a current generator. The input voltage V_x applied to the converter is transformed into a current that is integrated. The output voltage resulting from the integration has a slope proportional to V_x and of opposite sign. This signal V_A is applied to two comparators (1 and 2) which also have on their other inputs two reference voltages $+V_{ref}$ and $-V_{ref}$. Let us assume that V_x is positive. The output signal of the integrator decreases linearly and when it reaches the value $-V_{ref}$ comparator 1 changes state and generates one pulse. This pulse is counted by a counter and controls a switch which allows the current generator to send a calibrated positive current pulse to the summing junction of the integrator. This current pulse is used to discharge the capacitor and thus make it possible for a new integration to be carried out. Figure 65(b) shows the signals at various points of the circuit for two different values of V_x. To each integration corresponds a pulse and by counting these pulses during a given time (a second for example) an encoded representation of voltage V_x is obtained. When the voltage V_x is low the integration takes longer and the number of pulses decreases. The second comparator which is fed by the reference voltage $+V_{ref}$ allows the conversion of negative voltages to be carried out which gives rise to positive ramps.

The important factor to ensure good accuracy is the *current generator*. With properly designed generators an accuracy of 10^{-5} can be attained. In common with analog systems, this is slow since it involves counting a large number of pulses (2^n for a 1 second count). Under these conditions a stable integrator must be used which has no drift during the period of time of this count. This system has good noise rejection particularly as regards mains supply noise. Its noise performance is the same as that of the dual slope converter, since both converters carry out an integration operation. Because of its very high accuracy and its good noise rejection this system is used in high precision digital voltmeters.

Another principle used in V–F converters is shown schematically in Fig. 66.[36] In this method the problem is to count a number of pulses N, proportional to

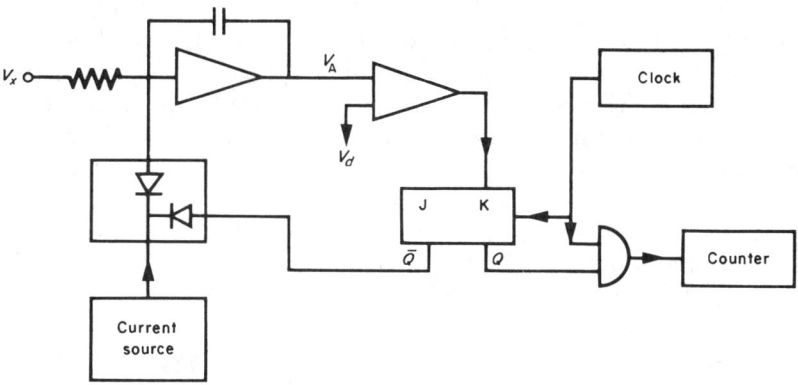

Figure 66

signal V_x, out of N_t pulses. The integrator, comparator, current switch and current generator are met again but the system also makes use of a clock and a logic circuit shown schematically by a flip-flop JK. The conversion operation comprises two stages:

when the output of the comparator is in state 1, the clock pulses are fed to the counter and the current source is connected to the integrator;
when the output of the comparator is in the state 0, pulses do not reach the counter and the current source is disconnected.

At the start of the conversion operation the unknown signal V_x is transformed into a current and then integrated, while the comparator is in the zero state. The counter does not receive pulses. When the output signal V_A of the integrator passes through the threshold value V_d set by the comparator, it changes state. The first stage of the operation is then initiated when the clock pulses are counted and a new current is integrated corresponding to the sum of the unknown current and the reference current supplied by the source. When the signal V_A passes through the threshold value again the count is stopped and the current source is disconnected. The time between two consecutive passages through the threshold value depends upon the slope of voltage V_A (hence of voltage V_x). During each operating cycle the counter adds a number of pulses related to V_x. A conversion cycle consists of N_t clock periods and during the corresponding time a certain number of operating cycles are carried out and only N pulses among the N_t possible pulses have been counted. This number N is related to the number of integration cycles. Although this number of cycles is not directly counted and a pulse is not produced at each integration, a number N proportional to V_x is obtained which can be considered to be the average signal frequency. If the threshold voltage V is made equal to zero then positive or negative voltages can be converted.

3.5.6 Capacitive Charge Transfer Converters

Another type of ADC can be classed among the analog ADC category; these are the *cyclic* or *capacitive charge transfer converters*.[39] The principle, which because of its repetitive nature is similar to that of the successive approximation conversion, consists in using the *transfer of charge* between two capacitors, hence the name given to that converter.

The system generates reference voltages which form a geometric progression of common ratio 2, by distributing a charge Q between two equal capacitors. The basic circuit diagram for generating the reference voltages is shown in Fig. 67.

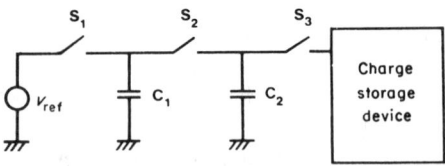

Figure 67

Capacitors C_1 and C_2 usually have the same capacitance. The charge storage device has the property that it will remove all the charge from C_2 when switch S_3 is closed and in a way serves as an integrator. The process is as follows:

C_1 is charged to the potential V_{ref} (S_1 closed, S_2 opened).
S_1 is opened and S_2 closed. C_1 and C_2 are charged to the potential $V_{ref}/2$.
S_3 is closed and S_2 opened. C_2 is completely discharged and the increment of
    voltage $V_{ref}/2$ is transferred in the charge storage device.
S_2 is closed and S_3 opened. C_2 is charged to a potential $V_{ref}/4$.
S_2 is opened and S_3 closed. The charge on C_2 is transferred.

Thus reference voltages are generated whose values vary in decreasing order of powers of two. Switch S_3 can be replaced by a changeover switch so that the polarity can be reversed during the transfer of the charge of C_2 according to whether the stored charge is to be increased or decreased. The timing diagram of such a system is shown in Fig. 68.

A standard way of implementing a system for storing the charges of capacitor C_2 is shown in Fig. 69(a). An amplifier is used with a feedback loop through a

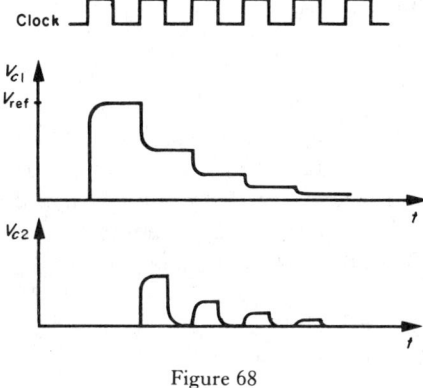

Figure 68

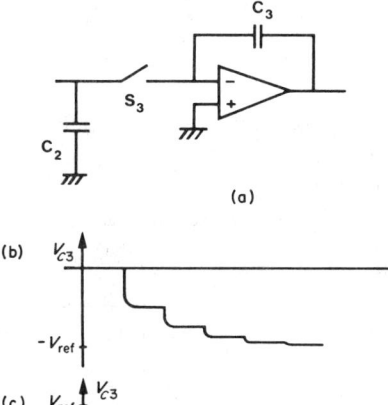

Figure 69

capacitor C_3 which provides a virtual earth at the amplifier input, so when switch S_3 is closed capacitor C_2 must discharge. Thus it supplies a current I which will charge capacitor C_3. The resultant change of charge ΔQ_3 is equal and opposite to the change of charge ΔC_2. If capacitor C_2 is completely discharged then:

$$\Delta Q_3 = -Q_2$$

If it is assumed that capacitor C_3 had an initial charge Q_{30} then:

$$Q_3 = Q_{30} - Q_2$$

and the voltage across the terminals of C_3 is given by:

$$V_{C3} = \frac{Q_3}{C_3} = \frac{Q_{30}}{C_3} - \frac{Q_2}{C_3}$$

If the two capacitors have the same capacitance, which is usually the case then:

$$V_3 = V_{30} - V_2$$

When capacitor C_1 is discharged in steps onto C_2 the capacitive charge of C_3 increases in steps. From the shape of voltage V_{C2} plotted in Fig. 68, the changes of voltage V_{C3} can be drawn (Fig. 69(b)). This voltage steadily increases at each clock period reaching successively the values $V_{ref}/2$, $3V_{ref}/4$, $7V_{ref}/8$ etc. If capacitor C_3 had an initial charge corresponding to a voltage V_i, the resulting voltage would be obtained by subtracting successively $V_{ref}/2$, $V_{ref}/4$, $V_{ref}/8$ etc., from V_i (Fig. 69(c)).

In order to apply the principle of charge transfer to the measurement of a voltage V_x, capacitor C_3 is charged by means of that voltage and the charge of C_3 is brought down in steps to zero by adding or subtracting to it the reference voltages $V_{ref}/2$, $V_{ref}/4$, $V_{ref}/8$, etc. This method has a number of advantages inherent to repetitive systems:

(1) the number of precision components used is small and there are only three capacitors,
(2) the same components are used to determine each bit,
(3) the binary sequence can be transmitted as soon as it is generated and there is no need of a buffer memory, and
(4) the number of bits can be increased by simply increasing the conversion time.

An example of implementation given in Fig. 70 and Fig. 71 shows the corresponding timing diagram. Capacitors C_1 and C_2 are used to obtain the required voltages $V_{ref}/2$, $V_{ref}/4$ etc., which are then combined with the voltage V_x stored on capacitor C_3.

In order to carry out the various transfer of charges the device uses seven switches whose function can be described thus:

(1) Voltage V_x charges capacitor C_2 via S_1.
(2) When S_2 is closed, C_1 is charged to the voltage V_{ref}.
(3) When S_3 is closed, charges C_1 and C_2 become equal.

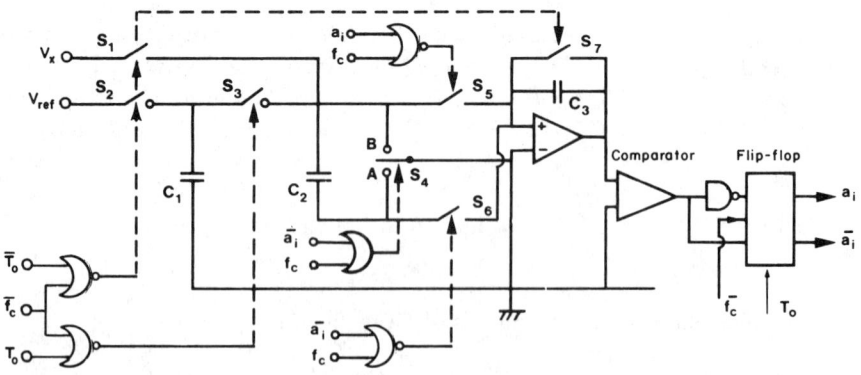

Figure 70

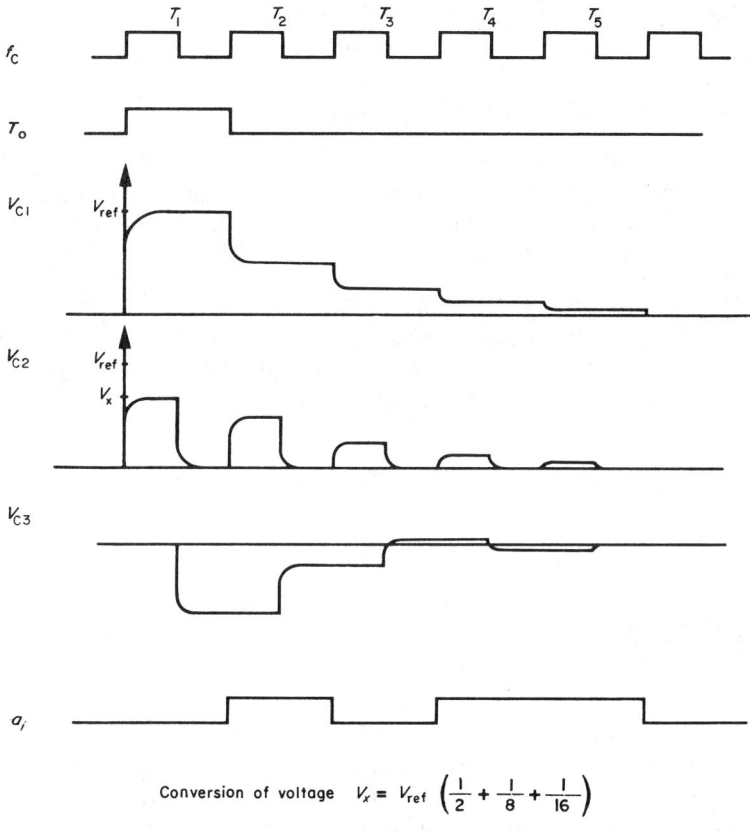

Conversion of voltage $V_x = V_{ref} \left(\dfrac{1}{2} + \dfrac{1}{8} + \dfrac{1}{16} \right)$

Figure 71

(4) S_4 earths one terminal of C_2.

(5) S_5 and S_6 connect one terminal of C_2 to the amplifier input.

(6) The state of output signal a_i determines whether S_5 or S_6 must be closed. If, during a period T_i, the voltage across the terminals of C_3 is negative, a_i is 1 and S_6 is then closed and voltage V_{C2} is added to voltage V_{C3}. If V_{C3} is positive, a_i is zero and voltage V_{C2} is subtracted from V_{C3}.

(7) Switches S_5 and S_6 can only close during the second half of each clock period (so as to have time to determine the sign of V_{C3}). All other switches close during the first half of each clock period.

(8) During the first clock period of a conversion operation a synchronization signal T_0 is generated in order to control the operation of certain switches. The operation of the switches can be expressed by logic equations. The control signals for the switches are as follows:

for S_1, S_2 and S_7 $f_c T_0$

for S_3 $f_c \bar{T}_0$

for S_{4A} $f_c + \bar{f}_c \bar{a}_i = f_c + \bar{a}_i$

for S_5 $\bar{f}_c \bar{a}_i$

for S_6 $\bar{f}_c a_i$

where f_c is the clock signal and a_i the ith bit.

This conversion system requires one period per bit and the first bit gives the sign of V_x. It is therefore a repetitive converter. Moreover it is bipolar since weighted voltages can be added to or substracted from V_x. The performance of such a system depends upon:

the time required to carry out the charge transfers,
the offset and gain errors of the amplifier and of the comparator,
the values of the capacitors and of the switch capacities. In order to minimize those effects it is necessary to operate at rather slow speeds using capacitors with high capacitance.

In this way an accuracy of 10^{-3} can be obtained for maximum conversion frequencies of about 20 kHz.

3.6 LOGIC ADCs

According to the classification we have adopted, a logic ADC^{41} is a converter in which the logic part is more important than the analog part (as far as the operation is concerned and often the implementation as well). The best known are the *parallel converter* and the *successive approximation* converter. Very fast converters such as *serial–parallel converters* could also be included but as they very often make use of special techniques they will be treated in a separate chapter. Thanks to the progress made by integrated circuit technology it is possible nowadays to manufacture an ADC in a single integrated circuit package.

3.6.1 Parallel ADCs

The parallel converter is a *multiple threshold converter*. The voltage V_x to be converted is simultaneously compared to $2^n - 1$ reference voltages having respectively the values:

$$\frac{1}{2}\frac{V_{ref}}{2^n}, \; \frac{3}{2}\frac{V_{ref}}{2^n}, \; \frac{5}{2}\frac{V_{ref}}{2^n}, \ldots, \frac{2^n - 3}{2}\frac{V_{ref}}{2^n}.$$

Thus the value of the number N corresponding to the voltage V_x is immediately found from among the 2^n possible combinations it can have. The parallel converter has therefore the following components:

$2^n - 1$ *comparators* each receiving two analog voltages; the voltage V_x to be converted and one of the $2^n - 1$ reference voltages.

A *resistance network* which will provide these voltages from a voltage V_{ref}. The precision of these resistors will depend upon the required number of bits.
A *logic circuit* which derives the number N in binary form from the comparators signals.

Figure 72 shows the basic schematic diagram of a three bit parallel ADC and the agreement existing between the digital word N obtained and the value of V_x relative to the various reference voltages.

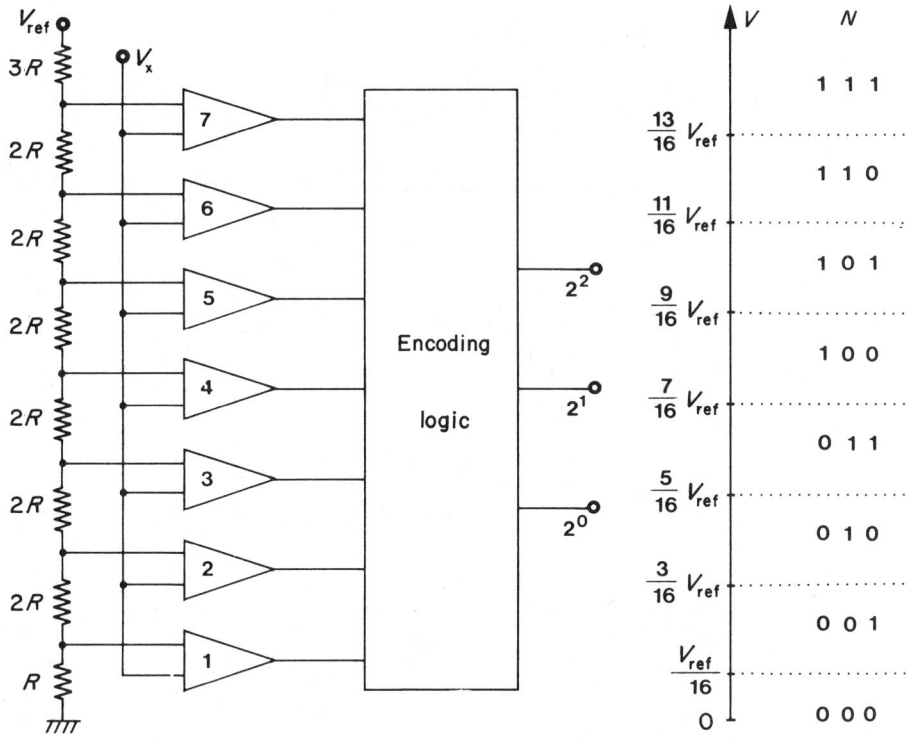

Figure 72

The operation of this system can be easily explained using Fig. 72. When the voltage V_x is applied to the converter input the comparators immediately determine whether this voltage is greater or smaller than the reference voltage which is also applied to them. The comparators can therefore be divided into two groups, the first p comparators for which V_x is greater than their reference voltage and whose output state becomes 1, and the remaining $2^n - 1$ whose output state stays 0. From this information the logic encoding circuit derives the digital word N. In order to carry out this encoding logic, the logic equations relating the bits of N to the state of each comparator must be written down.

In the case of the 3 bit converter of Fig. 72, if S_1 is the logic state of comparator of rank $i(1 \le i \le 7)$ and N_j is the value of the bit of rank j of the binary word

$N(0 \le j \le 2)$, then the following logic equations hold:

$$N_2 = S_4$$

$$N_1 = S_6 \vee S_2 \cdot \bar{S}_4$$

$$N_0 = S_1 \cdot S_2 \vee S_3 \cdot \bar{S}_4 \vee S_5 \cdot \bar{S}_6 \vee S_7$$

$\vee$ being the symbol for the logic operation OR. There are several ways of synthesizing these logic relations and Fig. 73 shows a logic diagram using only

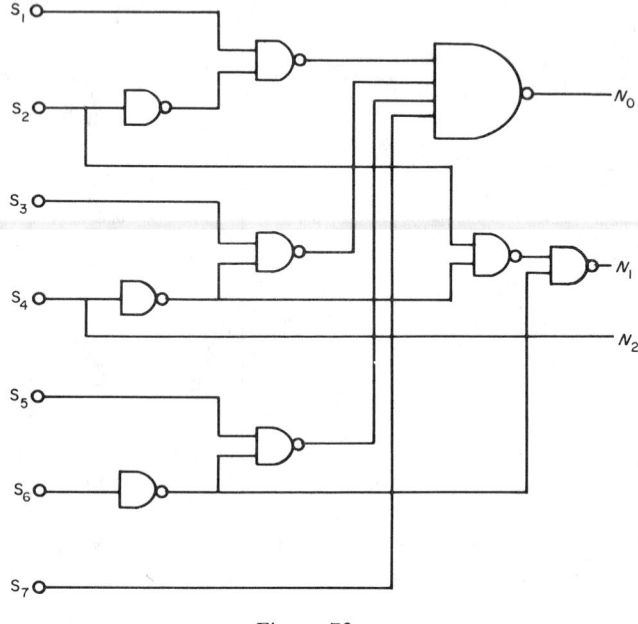

Figure 73

NAND circuits. The main advantage of this converter is its *very high speed*. The conversion times includes the time taken for one comparator to change state and the time required by the encoding logic to derive the various bits. (This time corresponds to the time taken by the signals to travel through the gates of the encoding logic.) By choosing appropriate comparators and using parallel encoding logic it is possible to obtain conversion times of the order of 10 ns.

This is the *fastest converter in existence*, but the speed is usually achieved at the cost of a high degree of complexity, for the number of components required increases in a geometric progression of common ratio two when the number of bits increases. Thus if $n = 7$, which gives an accuracy of 1%, it is necessary to have 127 comparators, a network of 128 resistors in order to obtain the reference voltage plus a substantial logic circuitry. The use of this converter is therefore limited to cases where a low resolution is adequate (4 to 5 bits) and where the required feature is speed.

For this converter more than for any other type the choice of comparator to be used is very important for it can limit the maximum conversion frequency. As seen by the signal generator driving it, each converter is equivalent to a parallel RC circuit. For an n-bit converter (comprising $2^n - 1$ comparators) the equivalent circuit seen by the generator consists of a capacitor of $(2^n - 1)C$ in parallel with a resistor $R/(2^n - 1)$. If R_g is the internal resistance of the generator then it is seen that the voltage applied to the comparator due to the generator is not V_x but:

$$V_x \frac{\left(\dfrac{R}{2^n - 1}\right)}{\left(R_g + \dfrac{R}{2^n - 1}\right)}$$

This potentiometer effect could be significant. Moreover this voltage is attained with a time constant nearly equal to $(2^n - 1)R_g C$. If no precautions are taken, or if n is too high, this time constant can become greater than the conversion time required. The accuracy of this ADC is determined by the comparators and the resistance network for obtaining the reference voltages. To the extent that the number of bits is kept low (for example five) the corresponding accuracy (3%) is obtained with no difficulty.

3.6.2 Successive Approximation Converter

This is the most commonly used ADC because of its good performance and relatively low price. It is to be found in most manufacturers' catalogues and will now be examined in detail. It is also sometimes called a *weighing converter* since its operating principle recalls the process of weighing by successive approximation using a set of weights. When a weight X less than 1 kg is to be weighed it is first compared with a weight of 500 g. If it is heavier, then a weight of 200 g is added. If it is lighter then the 500 g is removed and the 200 g weight is tried and so on until X is weighed with the required accuracy.

The successive approximation converter[40] principle is based on the same idea. It is a question of determining, one after the other, the values of the various bits of the binary word representing V_x starting with the MSB. To this end the following expression for V_x is used:

$$V_x = V_{ref}\left(\frac{b_1}{2} + \frac{b_2}{2^2} + \ldots + \frac{b_n}{2^n}\right)$$

The operation proceeds as follows:

In stage one, V_x is compared to $V_1 = V_{ref}/2$; if V_x is greater than V_1 then $b_1 = 1$ and $V_{ref}/4$ must be added; if V_x is less than V_1 then $b_1 = 0$ and $V_{ref}/2$ is replaced by $V_{ref}/4$.

In stage two, V_x is compared to $V_{ref}/4$ or $3V_{ref}/4$, that is to $V_2 = b_1(V_{ref}/2 + V_{ref}/4)$ depending upon the result of the previous comparison. If V_x

is greater than V_2, then $b_2 = 1$ and $V_{ref}/8$ must be added. If V_x is less than V_2, $b_2 = 0$ and $V_{ref}/4$ is replaced by $V_{ref}/8$.

The process is repeated with all the various voltages $V_{ref}/8$, $V_{ref}/16$, ... up to $V_{ref}/2^n$.

It can be stated after the last comparison, when $V_{ref}/2^n$ has been tried, that:

$$V_x - V_{ref}\left(\frac{b_1}{2} + \frac{b_2}{2} + \ldots \frac{b_n}{2^n}\right) < \frac{V_{ref}}{2^n}$$

The sum of the weighted voltage obtained represents the nearest approximation of V_x taking into account the accuracy required.

An analysis of the various stages of the conversion process reveals the three essential functional elements of a successive approximation converter:

Weighted voltages must be generated, and to this end a DAC is used.
It is necessary to compare two voltages so a comparator is required.
The output signal from the comparator must be processed before it can be used for the control of the DAC. This is achieved by means of a command and control logic.

Figure 74 shows a functional block diagram of such a converter and Fig. 75 shows the various possible changes of the output voltage V of the DAC during a

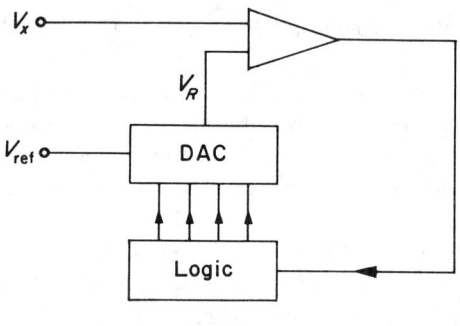

Figure 74

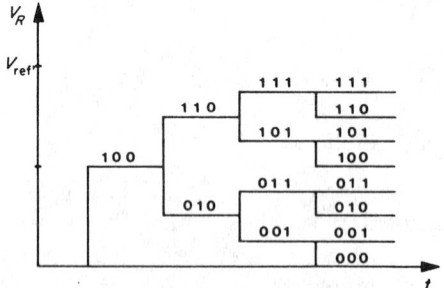

Figure 75

conversion operation according to the value of voltage V_x and for a three bit word. In practice the changes occur for odd multiples of $q/2$. The expression for the control signal of the DAC at each conversion stage is also shown and at the last conversion stage it is equal to the digital equivalent word N of the input voltage V_x. A successive approximation ADC is a *serial* (or sequential) ADC since one bit is obtained at each stage. Therefore n periods are needed for an accuracy of n bits. Each period is subdivided into two parts; firstly a comparison is made and then depending upon the result of that comparison the value of the bit in question is determined.

ADCs may be divided into *presubtractive* and *postsubtractive*. This distinction is now going to be explained for the case of a three bit converter and assuming that $V_{ref} = 8V$ and $V_x = 5.1$ V (which corresponds to $N = 101$).

In a presubtractive converter the voltage V_x is compared to the voltage $V = 4$ V, which gives $b_1 = 1$; then the voltage $V_2 = 4 + 2 = 6$ V is tried and $b_2 = 0(V_x < V_2)$; voltage $V_{ref}/4 = 2$ V is removed and replaced by the voltage $V_{ref}/8 = 1$ V (Fig. 76(a)). This means that in the case of the comparison being negative the last reference voltage added has first to be removed (in this case $V_{ref}/4$) before the next reference voltage can be added.

In the case of a postsubtractive converter when a comparison is negative the last reference voltage added is kept on and the next reference voltage is then subtracted. In the example, having established that $V_x < V_2$ a voltage V_3 is

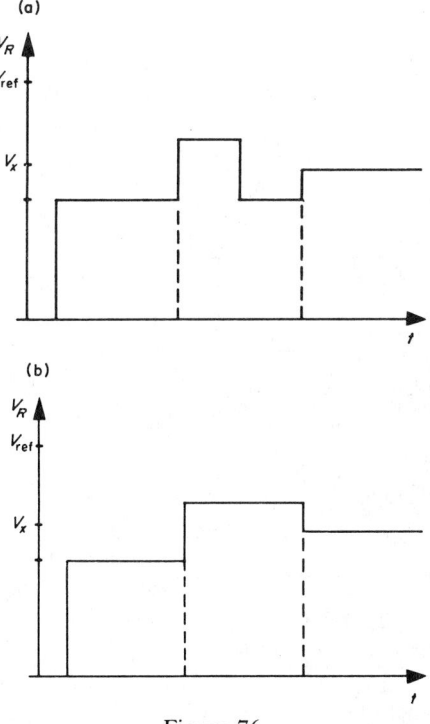

Figure 76

obtained such that

$$V_3 = \frac{V_{ref}}{2} + \frac{V_{ref}}{4} - \frac{V_{ref}}{8} \quad (\text{Fig. 76(b)})$$

In these circumstances it must be possible to add and subtract voltages whereas in a presubtractive converter only additions are catered for.

However these two systems are very similar and only the command logic is different since the algorithm used is not the same. Only the presubtractive arrangement shall be dealt with in what follows.

3.6.3 Organization and Operation

Because of the continuous development occurring in the design of various logic integrated circuit families it is difficult to claim that a particular logic circuit is the best. In fact a number of factors help decide whether a particular logic circuitry is suitable for a particular application. For example, it may be necessary to minimize the number of circuits used or the power dissipation. However the schematic block diagram of Fig. 77 is typical of successive approximation ADCs currently being manufactured. The main components are:

a DAC having a number of bits equal to or greater than the number of bits required to represent the voltage V_x,

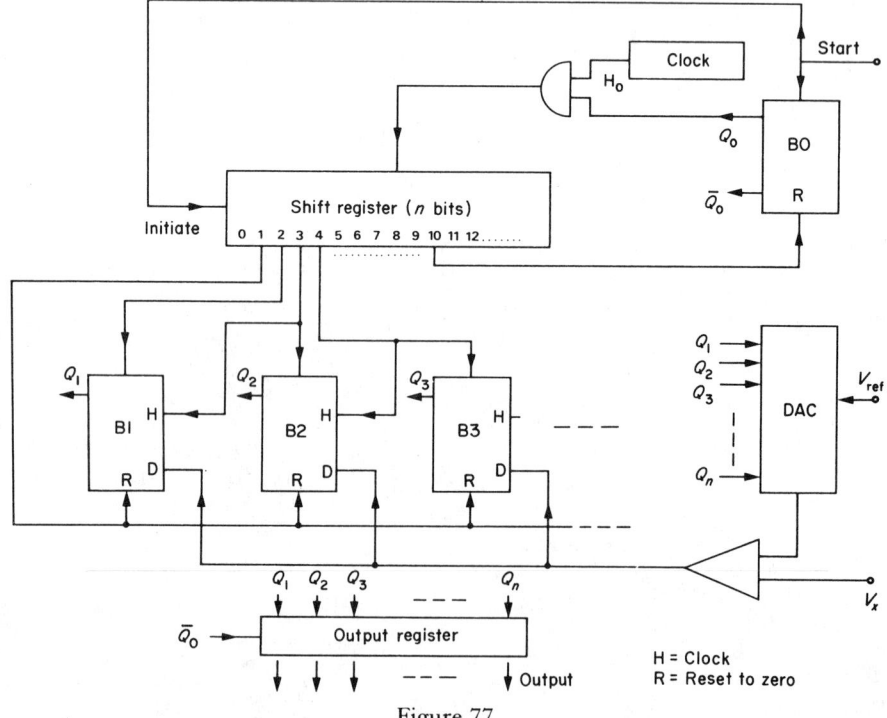

Figure 77

a *comparator* to which is applied both voltage V_x and the output voltage of the DAC,

a *shift register* which controls the generation of reference voltages in succession, that is, one after the other,

n *flip-flops*, B (one per bit), which give the values of the various bits of N. The output signals of flip-flops Q_1 to Q_n control the n inputs of the DAC,

an *output register* which stores the information as long as the conversion operation is not completed.

The operation of the system is now going to be examined. When a conversion operation is to be carried out a start signal is sent to the converter and applied to flip-flop B_0 which changes to the state 1 and allows the conversion to begin. The clock pulses are then fed to the shift register which controls the states of flip-flops B_1 to B_n. The start signal also sets the register. At this instant of time the output 0 of the register is in the state 1 and all the other outputs are in state 0. Each time the register receives a clock pulse the information held in the register is shifted one place to the right. Thus the various outputs will change to state 1, each one in turn, with all others remaining in state 0. In other words there is only one flip-flop driven at each clock pulse.

At the same instant of time the complementary signal $\bar{Q}_0$ supplied by flip-flop B_0 is sent to the output register so that the information being processed cannot be read out as long as the conversion operation is not completed. This signal $\bar{Q}_0$ remains until the flip-flop B_0 is reset to 0 and this only occurs when the last bit has been processed.

The state of the circuits used can change on the rising or falling of the clock signals and thus deliver positive or negative pulses. It is assumed here that the circuits change states on the rising edge of the clock signals and produce negative pulses. If this is not the case it would only be necessary to insert inverters at input D of the flip-flops.

The progress of a conversion cycle is as follows.

(1) When the first clock pulse is sent to the register, output one changes to state 1. The signal from this output resets to zero flip-flops B_1 to B_n. The input signal to the DAC is 000 . . . 00.

(2) The second clock pulse changes the state of output two to state 1. The signal from this output is applied to flip-flop B_1 and changes its state to 1 (a preset input is used for this). The input signal of the DAC is now 100 . . . 00 and hence applies to the comparator a voltage $V_{ref}/2$. If V_x is greater than $V_{ref}/2$ the comparator changes to state 1, if not it remains in state 0. Hence the value of b_1 is shown at the output of the comparator. This signal is sent to input D of the n flip-flops but cannot change their state since no signal is applied to the clock inputs of these flip-flops.

(3) The third pulse changes the state of output three to state 1. This signal puts flip-flop B_2 into state 1 (thanks to the preset input) and moreover it is sent to the clock input of flip-flop B_1 which reproduces the available information, that is the output state of the comparator, on its D input. In this way flip-flop B_1 must assume the state corresponding to the value of coefficient b_1.

Henceforth this state will not be changed, since no further signal shall be applied to the clock input of this flip-flop. The logic signal for b_1 applied to the DAC is then $10\dots00$ and the voltage that the DAC supplies is therefore $b_1(V_{ref}/2 + V_{ref}/4)$. The result of the comparison gives the coefficient b_2.

(4) At the fourth clock pulse the same operations are repeated with flip-flops B_2 and B_3. The output signal of the comparator is stored in flip-flop B_2 since it receives simultaneously a pulse on its clock input and the output signal of the comparator on its D input. Simultaneously flip-flop B_3 changes to state 1 and

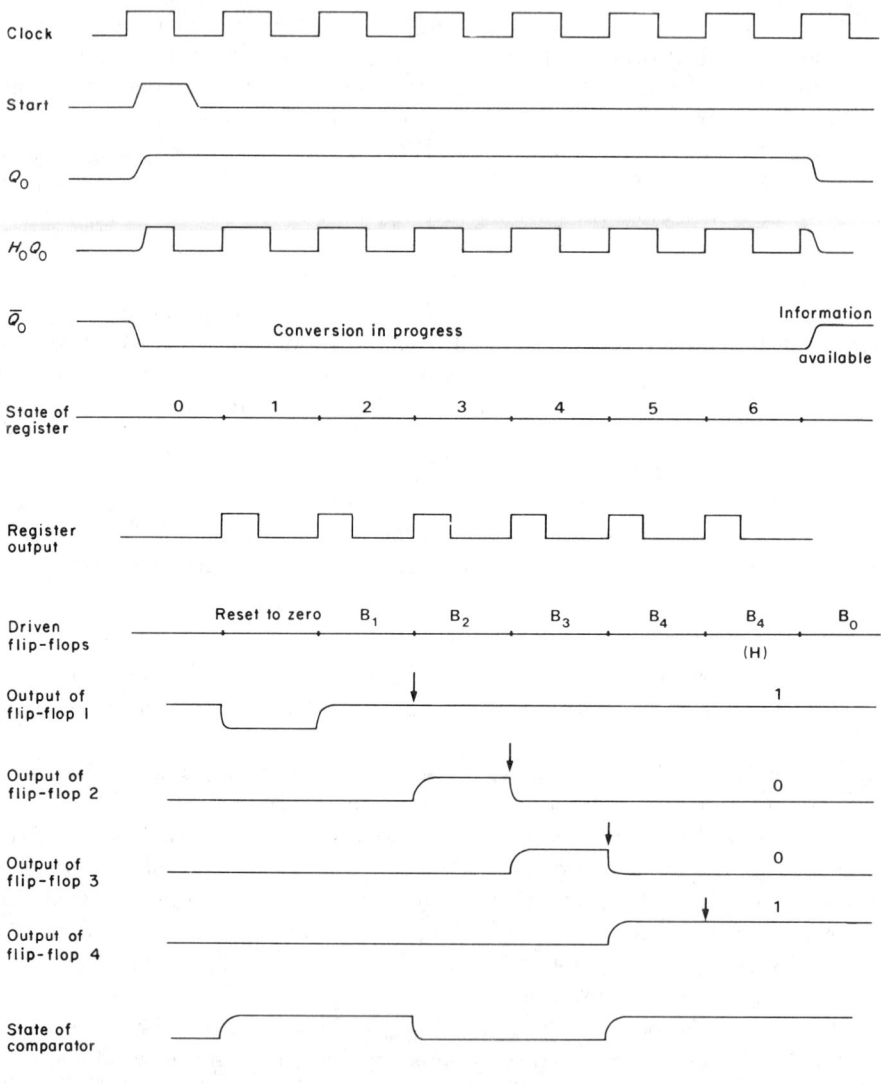

Conversion of a voltage equivalent to number 1001

Figure 78

the voltage output of the DAC is:

$$b_1 \frac{V_{\text{ref}}}{2} + b_2 \frac{V_{\text{ref}}}{4} + b_3 \frac{V_{\text{ref}}}{8}$$

There are therefore two operations to each conversion stage: checking the state of flip-flop B_{j-1} which had its state changed to 1 by pulse j, and changing to 1 the state of flip-flop B_j. This process is repeated until the nth flip-flop changes its state to 1 (with pulse $n + 1$). The coefficient b_n is then processed and pulse $n + 2$ is used to verify that coefficient. When pulse $n + 3$ is applied to the register its signal on output $n + 3$ resets flip-flop B_0 to zero. The conversion process stops and the coefficient stored in the output register become available. Figure 78 shows these steps for the case of a four bit word.

Depending upon the application requiring the converter, it can be advantageous to have the word N in serial or parallel form. When it is in *parallel* form the transmission is faster but requires n lines (one per bit). For a *serial transmission* only one line need be used but the time required is n times longer. In all converters the information N can be transmitted in serial form by transferring the word N into a buffer register at the end of the conversion and emptying it in serial form, though this increases slightly the complexity of the converter. This expedient is unnecessary with successive approximation ADCs since at each clock period one of the coefficients of N is available at the output of the comparator. It is only necessary to pick up this signal as it occurs during the conversion process to have the information N in serial form.

3.6.4 Performance

This converter is of great interest because of the *accuracy* obtainable and the *small operating time* required. It was shown when the operation of the converter was examined that in order to convert a voltage with an accuracy corresponding to n bits $n + 3$ clock periods were required. This time must be compared with the 2^n periods needed for a single slope converter. Therefore the successive approximation converter is a fast converter. The lower limit of the conversion time is determined by the time needed for the various operations that must be carried out during one period.

In particular, account must be taken of the *settling time* of the DAC, the *response time* of the comparator and the *delays* introduced by the various logic circuits. If it is required to increase the number of bits, the conversion time is only slightly increased since it is only necessary to add one clock period per additional bit (assuming that the times allocated to each coefficient do not depend upon the required accuracy which is not a very rigorous condition).

The second characteristic of interest is its accuracy, which is determined essentially by the comparator and the DAC. The DAC must have a number of bits greater than that of the ADC by at least one unit in order not to limit the accuracy ot the system. The effect of the comparator is more complex. When the number of bits is increased, the gain of the amplifier used in the comparator must also be increased in order that the comparator may detect that the differential voltage between its inputs has changed sign with the same relative accuracy (for example $\frac{1}{10}$ quantum). This usually results in a reduction of the slew rate. The characteristic of the comparator which is of interest is therefore the product gain-rate of change of the output signal. It is also useful to know the common mode voltage it accepts and the way it rejects it, the common mode rejection factor. The selection of a converter will be discussed in detail in a later chapter but by carefully selecting the comparator and the DAC it is possible to obtain an accuracy of 12 bits. If the DAC is not monotonic, certain combinations cannot be obtained at the output of the ADC and the differential linearity error would exceed one quantum. This converter can therefore be considered as providing one of the best compromises between accuracy and speed. It is not very complex and because of progress made by solid state technology it is now fabricated as a monolithic integrated circuit.

3.7 VERY HIGH SPEED ADCs

Very high speed ADCs[42] can be classified as either analog or logic ADCs since they often make use of the principles that have already been discussed. It has been thought best to treat them separately, since their use is rather specific. Up to now they have proved difficult to fabricate as monolithic circuits and because it is not easy to decide whether they ought to be considered as logic or analog ADCs.

Increase of conversion speed always leads to a significant increase in the complexity of the circuitry if the same accuracy is to be maintained. Moreover, the components must be selected appropriately and their response time must be very low. The starting point of the design of very high speed ADCs is to combine the advantages of the parallel converter and of the successive approximation converter and eliminate their disadvantages, so that the speed of the parallel ADC will be retained together with the moderate complexity and accuracy of the successive approximation ADC. Two methods can be considered for producing a very high speed ADC.

Carry out a *parallel–serial conversion*. In this case the various bits are produced in groups using several parallel ADCs which are connected in series. For example p ADCs are used each giving q bits so that $n = p \times q$.

A *variable threshold converter* can be used. This is a successive approximation ADC in which n reference voltages are generated by n DACs. The result of a comparison controls the DACs that follow and affects the values of the voltages used for the later comparisons.

As an example a serial–parallel converter will now be described. In this type of converter the bits are obtained q at a time and the operation is repeated p times to obtain the n bits required ($n = p \times q$). To this end p parallel ADCs each giving q

bits are used and the ADCs operate one after the other. Suppose that it is required to convert a voltage V_x into an equivalent word of six bits, the bits being obtained two at a time. The conversion requires therefore, three intermediate stages. A block diagram is shown in Fig. 79 and consists essentially of 3 parallel ADCs each supplying 2 bits, 2 DACs of 2 bits and 2 analog subtractor–multiplier modules.

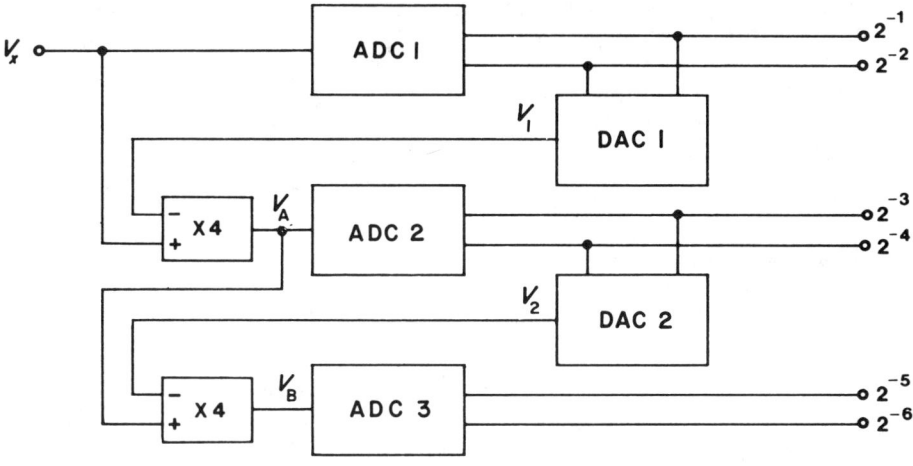

Figure 79

Each ADC has three comparators each having one of the reference voltages $3V_{ref}/4$, $V_{ref}/2$ or $V_{ref}/4$ applied to it. The DACs are fed by the same reference voltage as the ADCs. The analog modules subtract the input voltage of an ADC from the output voltage of the associated DAC and multiply the difference obtained by $2^2 = 4$. The various operating stages are as follows.

First ADC 1 determines the position of V_x with respect to three reference voltages that are supplied to it. It is then possible to write for V_x:

$$V_x = b_1 \frac{V_{ref}}{2} + b_2 \frac{V_{ref}}{4}$$

The ADC produces the two bits of N of greatest weight. These two coefficients are sent to DAC 1 which produces a discrete voltage V_1 which is a multiple of $V_{ref}/4$. It can be stated that the difference $(V_x - V_1)$ is less than $V_{ref}/4$.

Next this difference, multiplied by 4, is sent to ADC 2 which encodes it with 2 bits and thus

$$V_A = 4(V_x - V_1) = b_3 \frac{V_{ref}}{2} + b_4 \frac{V_{ref}}{4}$$

that is

$$V_x - V_1 = b_3 \frac{V_{ref}}{8} + b_4 \frac{V_{ref}}{16}$$

or

$$V_x = b_1 \frac{V_{ref}}{2} + b_2 \frac{V_{ref}}{4} + b_3 \frac{V_{ref}}{8} + b_4 \frac{V_{ref}}{16}$$

ADC 2 produces therefore bits 3 and 4 of N. These two bits are sent to DAC 2 which produces a voltage V_2.

Finally the operation is repeated taking now the difference $V_A - V_2$ and multiplying it by 4. The resulting voltage V_B is converted by ADC 3 thus:

$$V_B = 4(V_A - V_2) = b_5 \frac{V_{ref}}{2} + b_6 \frac{V_{ref}}{4}$$

$$V_A = 4(V_x - V_2) = V_2 + b_5 \frac{V_{ref}}{8} + b_5 \frac{V_{ref}}{16}$$

which gives for V_x

$$V_x = V_1 + \frac{V_2}{4} + b_5 \frac{V_{ref}}{32} + b_6 \frac{V_{ref}}{64}$$

These operations are summarized in Fig. 80 for the case of $N = 100111$. The operation of multiplication is equivalent to a magnification of the image by four.

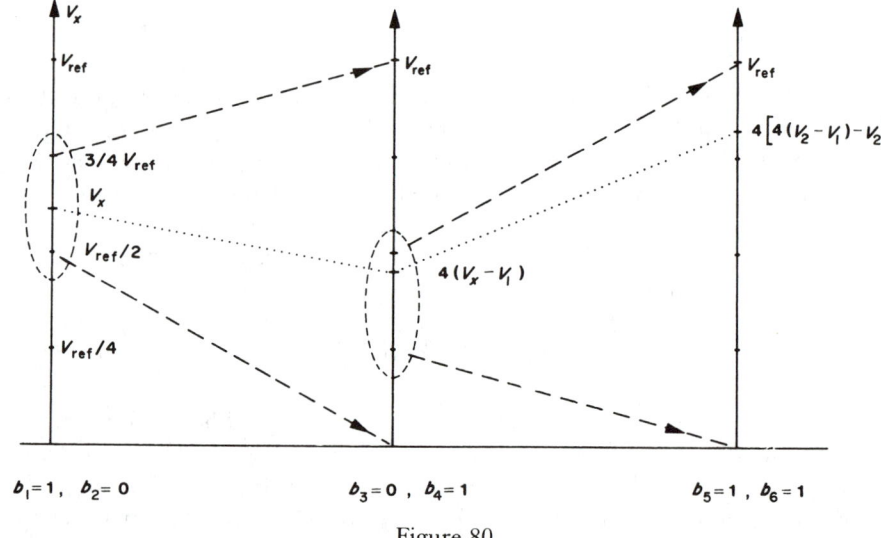

$b_1 = 1, \; b_2 = 0$ $b_3 = 0, \; b_4 = 1$ $b_5 = 1, \; b_6 = 1$

Figure 80

This converter is faster than a six bit serial converter but more complex. It is possible to produce the bits three at a time which reduces still more the conversion time at the price of greater complexity. A compromise must be reached between the time gained and the increase in complexity of the system when deciding how to apportion the n bits required. The accuracy of the system depends upon the intermediate ADCs and DACs used but they need not all have the same accuracy which in turn reduces the cost price. The overall accuracy depends mainly upon the first stage (ADC 1 and DAC 1). The error of the second stage (ADC 2 and DAC 2) can be four times greater than that of the first stage since its result is weighted by a factor of four. Similarly for the third stage the error can be 16 times worse and so on. It is therefore not worth using high precision components throughout.

3.8 BIPOLAR ADCs

In order to study the operation of the various ADCs it has been assumed for simplicity that all the signals to be converted were *unipolar* (positive). In practice signals are often *bipolar* and it is necessary to examine now if the various arrangements already discussed can be used with bipolar signals.[43] Some ADCs are inherently bipolar. This is the case of the single slope converter which delivers the word N equivalent to a voltage V_x in magnitude sign code. It is also the case for voltage to frequency converters in which an up–down counter arrangement is used to take care of the sign of V_x.

Where a converter is not inherently bipolar one solution consists in using a unipolar ADC preceded by *an 'adapter' circuit*. Figure 81 shows a connection used

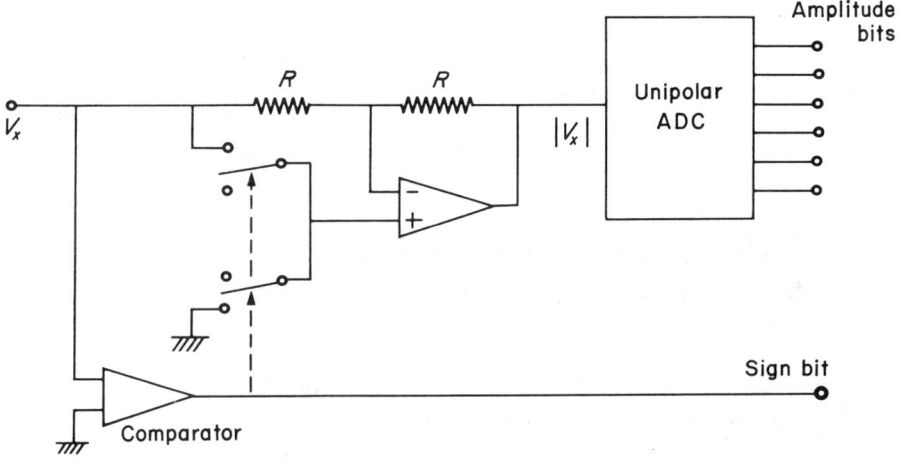

Figure 81

to carry out this adaptation. A comparator detects the sign of V_x and produces the sign bit. Its output signal controls a two-pole switch which is used to apply a signal equal to the absolute value of V_x to the input of the unipolar ADC.

A second solution consists in *modifying* or adapting the ADCs described previously. Many ADCs use a DAC to produce the reference voltages. To make such an ADC into a bipolar ADC only requires the use of a bipolar DAC which will produce voltages between the range $-V_{ref}$ and $+V_{ref}$. The dual slope ADC can easily be made suitable for bipolar signals. During the first integration the sign of V_x is determined and during the second operation stage a voltage equal to $|V_{ref}| \times$ (sign of $-V_x$) is integrated. Some ADCs generate themselves reference voltages from a voltage V_{ref} (using a divider circuit for example). In this case it is only necessary to use two voltages $+V_{ref}$ and $-V_{ref}$ in order to generate the reference voltages.

Finally it is sometimes advantageous to change from one bipolar code to another bipolar code. The logic statements that will be used to effect this change in code can be written with the help of Table 2 (p. 21). As an example Fig. 82 gives the block diagram for changing from an offset binary code to a magnitude–sign code for the case of a four bit word and using XOR circuits and a four bit adder.

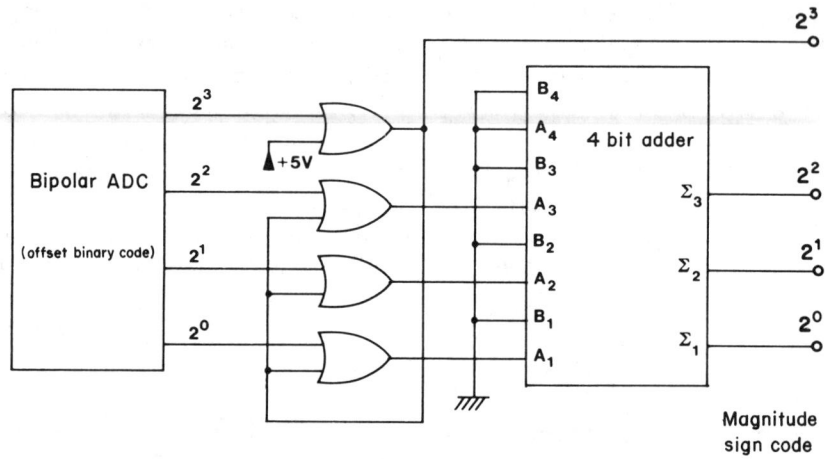

Figure 82

3.9 THE EFFECTS OF NOISE ON A–D CONVERSION

It is frequently thought that ADCs have an ideal transfer characteristic and the effects of *parasitic signals* are rarely analysed rigorously. However these disturbances considerably reduce the performance of high accuracy ADCs. Because of the inevitable presence of noise (inherent to the signal to be converted or arising from the components of the conversion circuitry) the signal produced by an ADC whose design has not taken noise into account, will contain a certain number of erroneous messages.[44,45]

In order to demonstrate the existence of errors in the messages produced by an ADC, an ideal (non-noisy) DAC is connected after the noisy ADC (that is which has its own sources of noise). The output message of the ADC is decoded by the DAC and its output voltage is compared to the input voltage of the ADC. In the absence of noise the difference observed corresponds to the quantization error but when noise is present this error is a function of the quantization error and of the error due to noise.

An ADC is said to deliver erroneous messages (or that its conversion is subject to errors) if the output number N' differs from the number N equivalent to the voltage V_x to be converted (taking into account the quantization interval), N and V_x being related by:

$$(N - \tfrac{1}{2}) \frac{V_{\text{ref}}}{2^n} < V_x < (N + \tfrac{1}{2}) \frac{V_{\text{ref}}}{2^n}$$

In order to know the probability of error, that is the confidence one can have in the result of a conversion operation, it is necessary first to examine the origin and nature of the noise and then analyse its effect on the encoding of analog signals.

The effects of noise are *statistical* in nature and can be described by the *probability* of a given error occurring at a given instant of time in the value of a signal. In digital transmission systems the effects of noise on the performance of the system can fairly easily be studied and the error determined as a function of the signal-to-noise ratio.

The phenomena occurring in ADCs are generally less well known. When a voltage supposedly constant and free of noise is converted, it is noted that the ADC produces two or more different messages. This is the *multistate* phenomenon which, however, is analogous to what occurs in a digital transmission. In a binary transmission there are only two possible states 0 and 1 and the errors correspond to the transposition of these two messages. On the other hand in an ADC which can supply 2^n different messages, the word N can become $N + 1$, $N + 2$, etc., or $N - 1$, $N - 2$, etc.

The noise due to the comparator will now be defined, then its influence on the message produced by the converter and the way in which it adds to the quantization noise will be examined.

3.9.1 Noise in the Comparator

The comparator is a fundamental component found in every ADC. Its operation becomes critical when high accuracy and high conversion speed are required. In order to be of high precision the comparator must have a large gain and this needs the offset voltages to be reduced and further that it should change state when a very weak differential signal is applied to it, which could perhaps be due to noise. In order to be fast it must have a wide pass band, and under these conditions of high gain and wide pass band the comparator is very sensitive to noise. It is therefore very important to know the influence of noise on the decision making process of the comparator.

The comparator is often the main source of noise in high accuracy ADCs. However, apart from the noise generated in the comparator itself there are other sources of noise which affect the decision. These are the noise in the internal resistances of the signal generator, the resistors of the ADC providing the weighted voltage, the noise from the mains supply etc. Changes also occur due to temperature or to the ageing of the components which can be considered as low frequency noise. It is possible to combine the noise external to the comparator, but affecting its decision, with its own noise. The appropriate sum of all noise components can be represented by one equivalent noise source at the input of the comparator which is itself then assumed to be noise free (Fig. 83). It is then

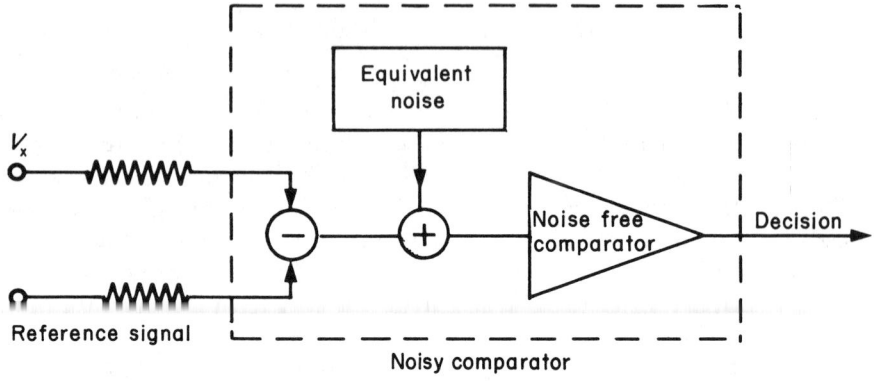

Figure 83

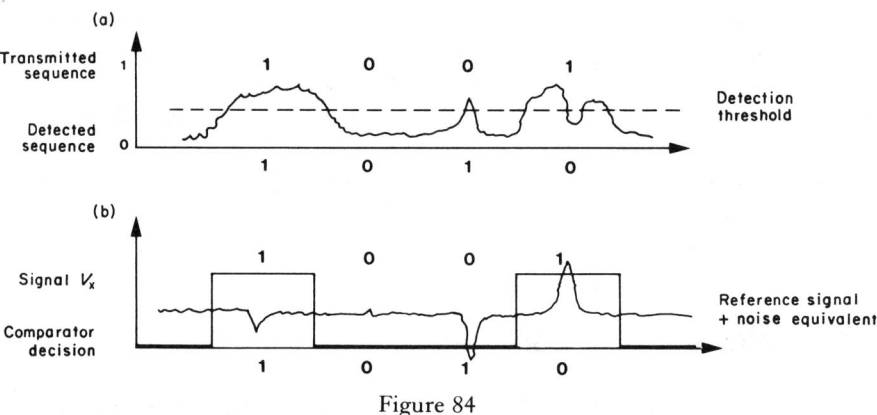

Figure 84

equivalent to apply to an assumed noise free comparator a noisy signal (Fig. 84(a)) or to apply a noise free signal to a comparator having a source of noise (Fig. 84(b)). The first representation is adopted for the analysis of digital transmissions. The second representation will be adopted for determining the influence of noise on the results of a conversion operation.

3.9.2   Noise Factor of a Converter

All ADCs have a source of noise that cannot be eliminated, the *thermal noise* due to the resistors. The total level of the noise present at the comparator stage will always exceed this noise by a certain factor and the evaluation of this factor will allow the quality of the circuit to be measured.

The noise generated in a resistive element is given by:

$$e_n^2 = 4kTR\Delta f$$

where e_n is the noise voltage expressed in rms volts, k is Boltzman's constant, R is the value of the resistance, T is the absolute temperature, and Δf is the pass band width. The relative contribution of the comparator noise can be expressed by means of the comparator noise factor F_C defined by:

$$F_C = 10 \log \frac{e_{ns}^2 + e_{nc}^2}{e_{ns}^2} \text{ dB}$$

where e_{ns} is the noise voltage due to the resistors external to the comparator and e_{nc} is the noise voltage generated by the comparator and referred to its input.

An overall noise factor F_{ADC} for the converter can also be defined taking into account all the sources of noise:

$$F_{ADC} = 10 \log \frac{e_{ns}^2 + e^2}{e_{ns}^2}$$

where e^2 is the sum of the noise of the comparator proper and of the noise originating from other sources and referred to the comparator input.

The overall noise factor of the converter usually ranges between 2 dB and 20 dB depending upon the quality of the components used. The number of errors due to the total noise referred to the comparator input depends upon the ratio of the amplitude of this noise to the amplitude of one quantization step of the comparator. In practice the full-scale of the comparator differs from that of the converter and this is due to many factors such as common mode rejection, etc. If V_C is the comparator full-scale, the value of one quantum for the comparator is given by:

$$q_C = \frac{V_C}{2^n}$$

If σ is the rms value of the comparator noise voltage then a factor K can be defined, $K = \sigma/q_C$ which allows the noisy feature of an ADC to be determined.

3.9.3   A Brief Note on Noise

In order to understand the effects that the relative amplitudes of the comparator noise and the quantization levels have on the converter performances it is necessary to recall very briefly certain noise characteristics. Noise is characterized by the *changes of its amplitudes* during a certain period of time, which can be described by stating the frequencies at which the various noise amplitudes occur. It is usually assumed that this distribution is *Gaussian* and satisfies the equation:

$$p(x)\, dx = \frac{1}{\sigma\sqrt{2\pi}}\, e^{-(x-\bar{x})^2/2\sigma^2}\, dx$$

p(x) dx is the probability of the amplitude of the noise lying between x and $x + dx$, $\bar{x}$ is the mean value of the amplitude and σ the standard deviation.

The distribution function can be made symmetrical about the origin (Fig. 85) which amounts to making $\bar{x} = 0$. The abscissa now represents the deviation with respect to the mean value expressed as a function of σ. The area under the curve gives the probability for the error not to exceed a certain value. There is always a probability, however small, for the occurrence of large amplitude noise.

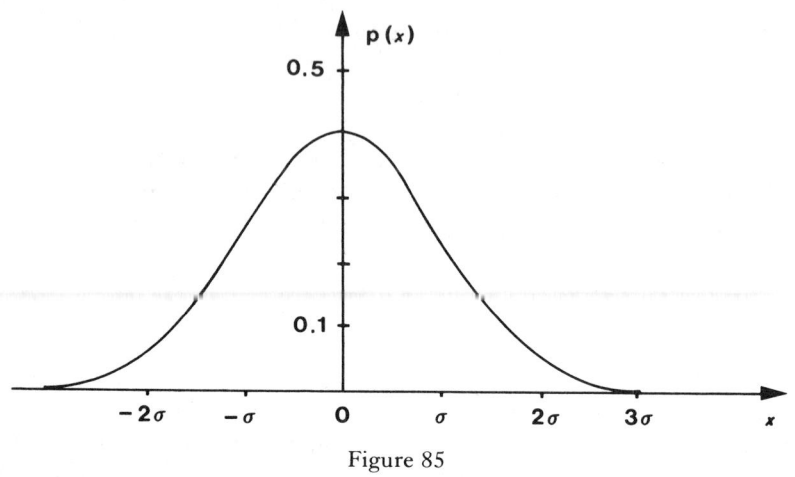

Figure 85

3.9.4 Encoding Errors Introduced by Noise

The error introduced when noise corresponding to the various values of the factor K is applied at the input of an ADC, assumed non-noisy, will now be examined. It will first be assumed that the input voltage is a whole multiple of the quantum, meaning that its defining point on the transfer characteristic is midway across a step (there is no quantization error). Figure 86(a) shows the positive distribution of the encoding errors (the conversion output number N' is greater than the actual number N) for various values of the factor K. The standard deviation remains the same and the quantum is referred to the standard deviation according to the value of K. Hence the probability of an error of one quantum, two quanta, etc., can be calculated. The table of Fig. 86(b) indicates the overall probability p(x) of an error, whatever its source, taking into account the value of K.

Now let us assume that the analog voltage to be converted is no longer placed symmetrically within the quantization interval but that it is shifted towards one end of that interval. The distribution curve of the amplitudes will also be offset since it is symmetrical about that voltage. Even if the noise voltage is very much less than the quantum the probability of an error greatly increases on one side (the side corresponding to the step towards which the voltage to be converted has been shifted) and diminishes a little on the other (Fig. 87). Compared with the previous case the probability of error will be higher, particularly for low values of K.

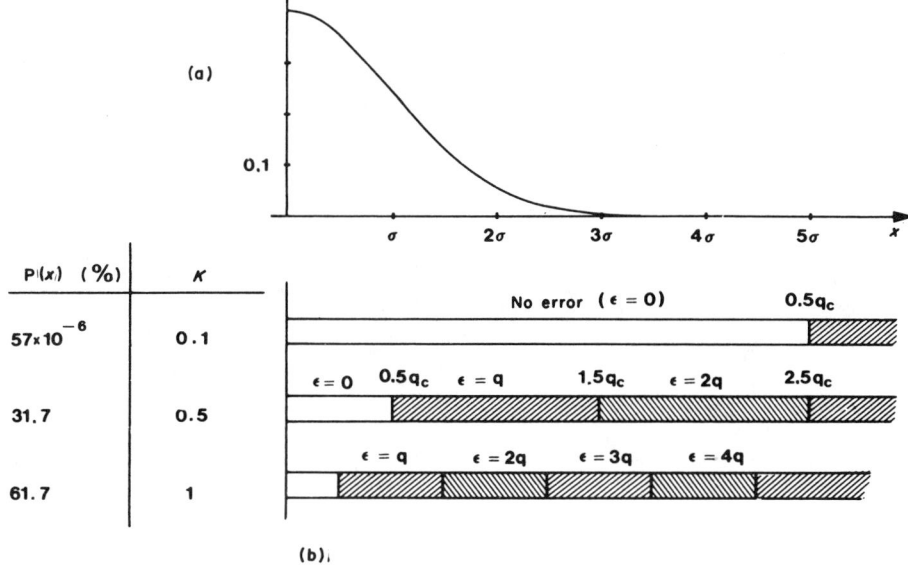

Figure 86

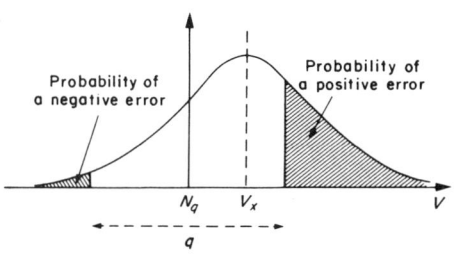

Figure 87

The curves of Fig. 88 show the probability of having at least one error for voltage V_x centred, or for V_x shifted with respect to the mid-point of the quantization interval. For a centred voltage the probability of an error is low if K is less than 0.2 (-14 dB); on the other hand it can practically never be neglected for an off centre voltage.

If the ADC is followed by an assumed noise free DAC in order to reproduce an analog voltage then it is of interest to be able to know *a priori* if the reconstituted voltage will be a good representation of the voltage V_x encoded by the ADC. The resulting error will depend upon the quantization error, whose mean quadratic value is $q^2/12$ and of the noise introduced by the ADC. Figure 89 shows the variation of the effective value of the restitution error η referred to quantum q as a function of the factor K. This error is adequately given by:

$$\eta^2 = \tfrac{1}{12} + K^2$$

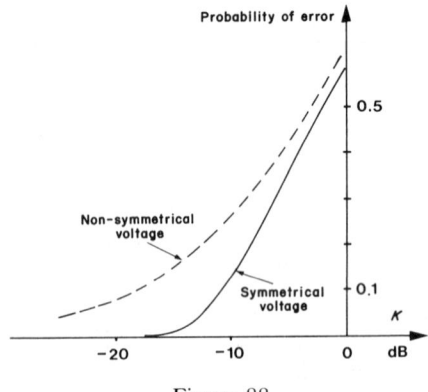

Figure 88

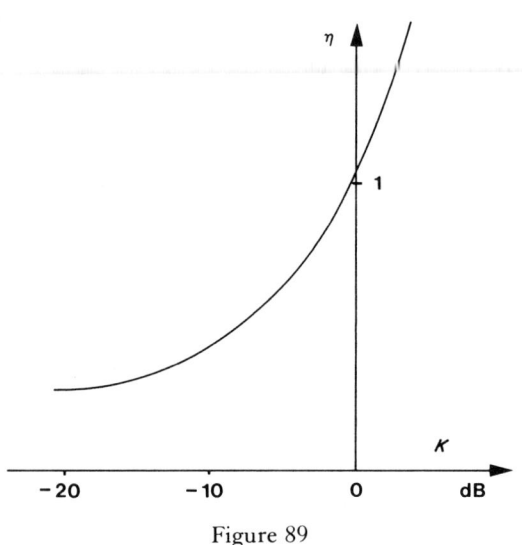

Figure 89

An example will clarify these concepts. A ten bit repetitive converter has a conversion frequency of 10 MHz. Full-scale for the comparator is 200 mV and the source impedance is 100 Ω. The input noise factor is 6 dB as is that of the comparator. What is the probability of error due to the first stage?

The value of factor K must be determined in order to make use of the curves of Fig. 88. For the comparator the quantum value is:

$$q_C = \frac{200 \, mV}{2^{10}} = 195 \, \mu V.$$

Now the noise voltage σ must be determined. The time allocated to one conversion is: $t = 10^{-7} \, s = 100$ ns. Each elementary step must be carried out in 10 ns. In order to obtain an accuracy of 10^{-3} (corresponding to ten bits) the time constant τ used must be:

$$\tau = \frac{10}{6.9} = 1.4 \text{ ns}$$

Assuming a first order circuit then the equivalent pass band is

$$\Delta f = \frac{1}{2\pi\tau} = 115 \text{ MHz}$$

The noise voltage due to the resistance of the source is:

$$e_n = 17 \text{ } \mu V_{rms}$$

The noise factor of the converter is therefore:

$$F_{ADC} = 6 \text{ dB} + 6 \text{ dB} = 12 \text{ dB or } F_{ADC} = 4$$

and the noise voltage σ is then:

$$\sigma = 17 \times 10^{-6} \times 4 = 68 \text{ } \mu V_{rms}$$

Hence the value of the factor K:

$$K = \frac{\sigma}{q_C} = \frac{68\sqrt{2}}{195} = 0.69 \quad \text{or 6 dB.}$$

If the voltage V is centred, the probability of an error occurring is 30%, if not it is 38%.

3.10   CRITERIA FOR TESTING AN ADC

Testing ADCs[29,46-48] is a more complex problem than testing DACs because of the unavoidable quantization error. In order to test an ADC it is necessary to know the output code and the transition points in relation to the input signal. Moreover the noise (whether it is carried by the signal, is produced in the converter or is picked up by the connecting leads) introduces an uncertainty into the accurate determination of those values of the analog input which produce a transition and results in an increase of the quantization range.

3.10.1   Calibration

Before testing an ADC the gain and offset errors must be eliminated. There are a number of preset converters produced for which the manufacturers do not provide access to the settings of the offset voltage and of the gain. However, when the required resolution exceeds eight bits it is often useful to have access to these settings. The *offset voltage* can be measured by observing the value of input voltage which produces the first transition. This transition must occur for a voltage of $\frac{1}{2}(V_{ref}/2^n)$. The difference between this value and the value that effectively triggers the transition corresponds to the offset voltage. Its effect is eliminated by adding to it an adjustable voltage of opposite sign. When all the bits are in the state 1 the output voltage is

$$V_{ref}\left(1 - \frac{1}{2^n}\right)$$

In fact it is useful to know the last transition. This must occur for a voltage equal to:

$$V_{ref}\left(1 - \frac{3}{2} \cdot \frac{1}{2^n}\right)$$

If it occurs for another value then the *gain* of the ADC must be adjusted. These two operations, that have to be repeated as many times as necessary, calibrate the ADC.

3.10.2 Testing an ADC

Having calibrated an ADC its *linearity* must next be checked. The simplest method consists in supplying the ADC with an accurate reference voltage checked by a digital voltmeter and displaying the result (Fig. 90(a)). The voltages of interest are those triggering the transition steps. A DAC may be used to produce the voltage for the ADC to convert, but it must have a higher resolution than that of the ADC being tested (Fig. 90(b)). An automatic system can control the DAC

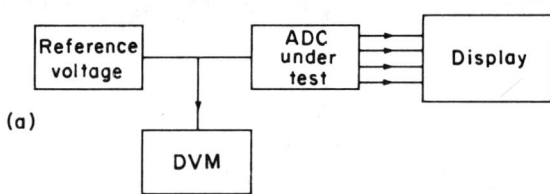

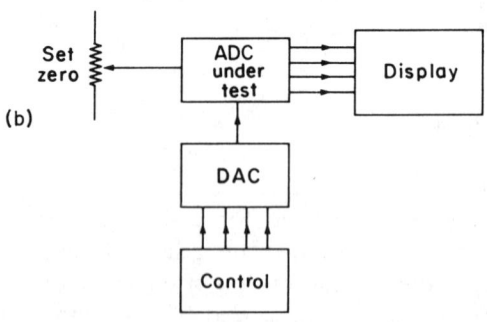

Figure 90

and check the 2^n values produced by the ADC (in this case only the nominal voltages can be checked without determining the transition points), and this method may also be used to check the *monotonicity* of the device. When using this method or the methods to be described below it is necessary that the accuracy of the voltage source used be greater than that of the ADC under test.

A second method consists in using a *precision DAC* to convert the output signal of the ADC and compare the input voltage with the reconstituted voltage (Fig. 91). This method may also be used to evaluate the differential linearity and can be

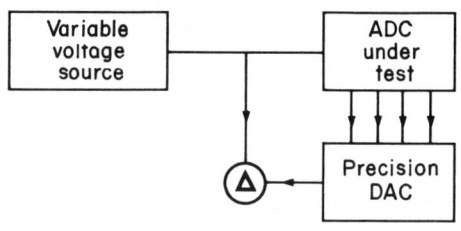

Figure 91

automated by using a low frequency generator supplying triangular waveform signals and a comparison circuit having a *window*. (In order that the results of the comparison be meaningful they have to be checked at well defined instants of time.) One must check that the error is in the window and the procedure must be fairly slow so that the transient conditions do not interfere. For an eight bit ADC a simple comparator can be used to compare the two voltages, but for a higher resolution ADC (12 bits) the comparator offsets and drift preclude its use, and a low-signal differential amplifier plug in unit from an oscilloscope can be used.

Another testing method consists in examining the trace shown on an oscilloscope by the output of the ADC whose input is controlled by a generator. Theoretically it is possible to display the whole transfer characteristic by stepping the sweep (or by using a double sweep). In order to avoid confusing the non-linearities of the generator driving the ADC with the linearity of the ADC, the same generator must be used for the horizontal sweep. In fact rather than sweep the whole range of the ADC it is preferable to apply a low amplitude voltage superimposed on a continuously adjustable DC voltage, varying it slowly if need be. By a proper choice of the amplitude of the alternating voltage (equal to $(4V_{ref}/2^n)$ for example) it can be arranged that only the last two bits change for a given value of the superimposed DC voltage. A two bit DAC implemented simply with two resistors R and $2R$ is adequate for displaying these four levels (Fig. 92). If the variable voltage is of the order of 1% of V_{ref} then there is a difficulty in defining a reference common to all the equipment. This can be resolved if the DC polarization source is not earthed. Figure 93 shows various oscillograms that may be obtained for particular non-linearities.

Another method consists in using a *reference ADC* (which has a higher resolution than that of the ADC to be tested). Both ADCs receive the same voltage to be converted and the two digital outputs are compared. To this end it is quite

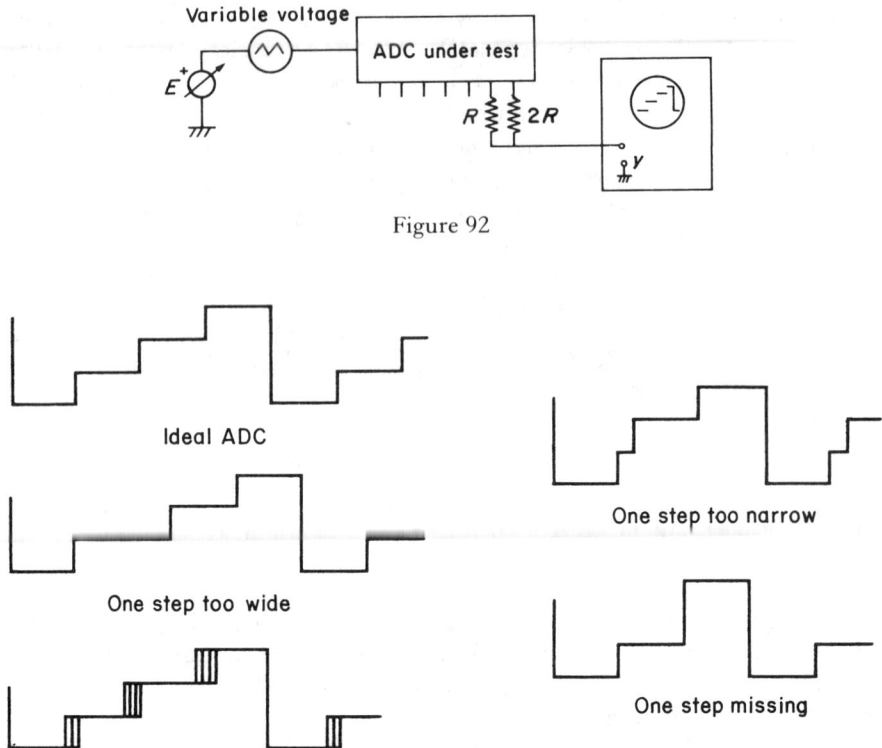

Figure 92

Ideal ADC

One step too wide

One step too narrow

One step missing

Presence of noise

Figure 93

adequate to carry out a straightforward bit by bit comparison by sending the two bits of the same weight to an exclusive OR circuit.

3.10.3 Statistical Test Methods

Statistical measurements are becoming increasingly important in areas such as real time adaptive encoding, optimal control or adaptive filtering. They require ADCs to encode the information and these ADCs must have very good linearity. Theoretically the width of all the steps of the transfer characteristic of an ADC are identical and each output word appears at the same frequency if all input amplitudes have the same probability of appearing. In practice the width of the steps is not constant and this results in errors in the frequency of appearance of output messages. Most current ADCs are likely to display this error. Thus an ADC with a feedback loop and a normal linearity error of $\pm\frac{1}{2}$ quantum results in a statistical error that can reach 50%. If the required statistical error must be less than 1% then only the two bits of greatest weight can be used for an eight bit ADC.

It is therefore necessary to specify a *new linearity criterion* enabling the linearity of an ADC to be tested. Its advantages will be as follows:

the measurement will be very sensitive (half the measurement scale corresponding to $\frac{1}{2}$ a quantum),
the result will give an exact value rather than a tolerance on the error,
the integration will eliminate the effect of noise, and
the analog part of the measurement will be simple and inexpensive and will not require precision equipment.

First the influence the linearity errors have on the statistical errors will be examined. Consider the case of a three bit ADC. If it were an ideal ADC all the steps would have the same width and, for an even distribution of the input signal, the probability of appearance of all output numbers is the same. If the transitions do not occur for the design voltages then errors occur. Suppose that the LSB is offset by $\frac{1}{2}$ quantum and all other bits are correct (Fig. 94). This produces two effects; the quantization error is no longer as specified ($\pm q/2$) and there is then a large variation in the frequency of appearance of certain output words when the distribution of the analog signal is uniform.

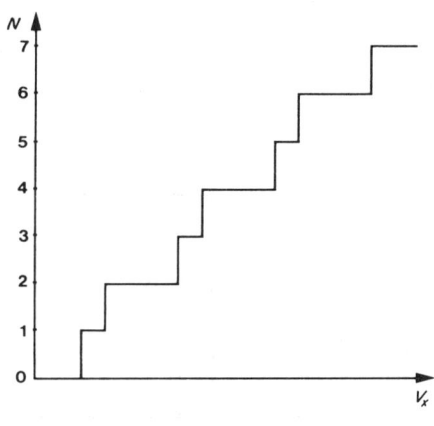

Figure 94

In the example of Fig. 94, the state 010 appears three times more often than the state 001. Thus even if the linearity error remains within the tolerance of $\pm q/2$ the statistical error is 50%. Each even number has a frequency of appearance of its own value of 150% and each odd number of only 50%. This results in a modulation (or variation) in the distribution of the probabilities of appearance such that this function can no longer be used to recognize the characteristics of certain signals.

The layout of a statistical test method for ADCs is shown in Fig. 95. At each clock pulse generated by the control module, the converter processes the signal supplied by the sawtooth generator, a signal having a uniform distribution, and converts it into a binary word. This word is stored in the address register of the

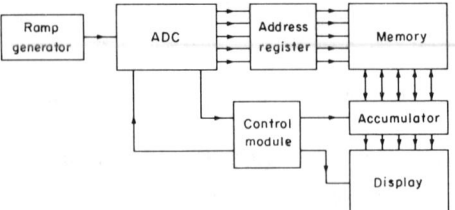

Figure 95

memory and indicates the memory location which registers the presence of this word and whose content is then increased by one. Thus the digital message serves only to indicate the address of a memory location in which the number of times the word appears at the output of the converter is to be stored. If one does not want to use too large memories then only the least significant bits are retained, for example the last six or last eight bits. This reduces the size of the memory required to 64 or 256 locations. At the completion of the test it is possible to display the result held in the memory. Thus the number of appearances of the various combinations is known and the linearity and monotonicity of the device can be deduced, the missing messages resulting in a 0. Thus in the case of Fig. 96 the word five cannot be obtained.

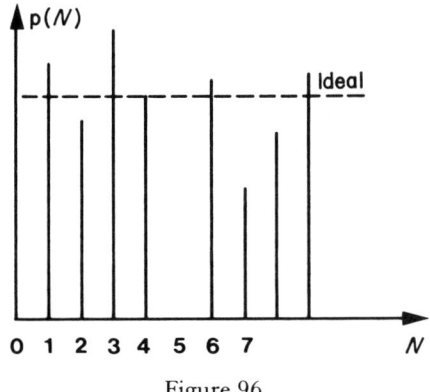

Figure 96

If a minicomputer is used then it is possible to carry out certain calculations with the information stored in the memory. The variance of the converter can for example be calculated and is given by:

$$\sigma^2 = \frac{1}{2^n - 1} \left[\sum_{j=1}^{2^n} (p(j) - \bar{p})^2 \right]$$

$p(j)$ being the probability of appearance of the word j and $\bar{p}$ the mean probability of appearance for an ideal converter. In this case it is important to know the minimum number of conversion operations that need to be carried out in order to have significant results. For a normal distribution curve the number of conversions required, N_0, is given by:

$$N_0 = 2^n \left[1 + \frac{1}{2} \left(\frac{Z_\alpha}{d} \right)^2 \right]$$

d is the accuracy with which the variance is to be known and Z_α is the abscissa of the normalized symmetrical normal distribution curve such that the area under the curves bounded by $-Z_\alpha$ and $+Z_\alpha$ is equal to the certainty with which the variance is to be known (Fig. 97). For example it is necessary to have $Z_\alpha = 1.65$ if a probability of 90% is required. In order to test a 7 bit ADC, 17 500 conversions must be carried out, and if the conversion frequency is 40 kHz the time needed is 0.5 s.

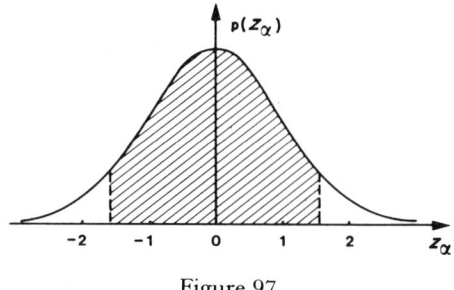

Figure 97

3.11 EXAMPLES

Having examined the operation of various ADCs it is of interest to compare their performances in order to be able to select an ADC for a particular application. The performances can easily be compared on *the maximum accuracy* expected, the *conversion time* and the *noise rejection*. The cost of the system is also a selection

Table 6. A comparison of the characteristics of the main types of ADC

Type of converter	Accuracy	Speed	Noise rejection
Single slope converter	medium	medium-slow	no
Dual slope converter	very high (10^{-5})	slow	yes
V–F converter	very high (10^{-6})	very slow	yes
Capacitive charge transfer converter	medium	medium	no
Parallel converter	low–medium	very high	no
Successive approximation converter	high	high	no
Serial parallel converter	medium	very high	no

criterion, though it is difficult to compare the prices of different system because on the one hand their performance can be so very different while on the other hand prices vary very rapidly with technological advances and with demand. Table 6 compares the characteristics of the main types of converters. The most accurate

Table 7. Characteristics of 3 ADCs that use different principles. Values at 25 °C if not stated

	Manufacturer (type)		
	Burr-Brown (ADC84 KG10)	Intersil Datel (ADC UH 8B)	Hybrid System (ADC 585)
Resolution (bits)	10	8	12
Principle used	successive approximations	series parallel conversion technique	double ramp
Input signal	±2.5 V; ±5 V; ±10 V 0 5 V; 0 10 V	±1.28 V 0 0.56 V	±10 V
Input impedance	direct: 0–5 V; ±2.5 V : 2.5 kΩ 0–10 V; ±5 V : 5 kΩ ±10 V : 10 kΩ with input amplifier: 100 MΩ	100 kΩ, 20 pF	0–10 V: 10 MΩ ±10 V : 400 kΩ
Output signal Unipolar Bipolar	complementary binary complementary offset binary complementary 2's complement series, parallel outputs	binary offset binary	binary offset binary series parallel outputs
Command signal for start of conversion	positive pulse of minimum width 50 ns TTL compatible	positive pulse of 2–5 V width 45 ± 5 ns	positive pulse, CMOS compatible
Conversion time (ns)	6	120	100
Accuracy (as % of full scale)		±0.4	±0.02
Temperature coefficient (ppm °C^{-1})	gain; ±30 accuracy; ±3 offset; unipolar: ±3 bipolar: ±15	overall coefficient ±50	linearity ±20 accuracy ±30
Supply	+15 V 45 mA −15 V 35 mA +5 V 70 mA	+15 V 80 mA −15 V 9 mA +5 V 1300 mA −5 V 250 mA	+15 V 30 mA
Sensitivity of supply	±15 V: ±0.004% of full scale per % +5 V: ±0.001% of full scale per %		±0.001% of full scale per %
Standard temperature range (°C)	0–70	0–70	0–70

systems are those with a feedback loop, the V–F converter or the dual slope converter and moreover these have good noise rejection. The highest speed converters are the parallel or serial–parallel systems but they have no noise rejection.

Table 7 reproduces for information the characteristics of three ADCs using different principles. Their characteristics being intentionally different no comparison should be made between them. Comparison should be made with converters having similar performance such as speed or accuracy for example.

3.12 LOGARITHMIC CONVERTERS

Analog to digital converters whose input–output function is *logarithmic* or *quasi-logarithmic* constitute a special class of converters of great interest to system and network engineers.[2,49–51] They can be used to solve numerous problems of data acquisition, encoding, or transmission when it is required to obtain a dynamic range of several decades with a constant *relative* accuracy (and not absolute accuracy). The choice of a quantization law is linked to the distribution of the input signal V_x. It often happens that the probability of appearance of the signal V_x is inversely proportional to its own amplitude and in this case the best quantization law is a logarithmic one. It is possible to classify the logarithmic ADCs (LADCs) into three categories according to the block diagram of Fig. 98.

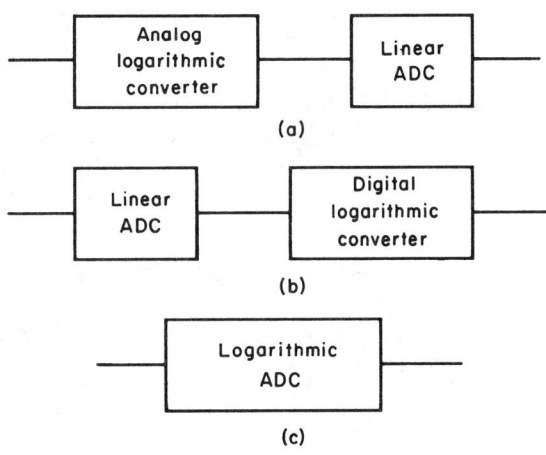

(a)

(b)

(c)

Figure 98

The LADCs grouped in the first category consist of a cascade arrangement of a *logarithmic converter* treating *analog variables* and supplying (outputing) analog quantities, and of a conventional linear ADC. In the second category the *linear ADC* is followed by a *digital logarithmic converter*. As for the third category the *same arrangement* carries out the analog to digital conversion and the logarithmic conversion without it being possible to separate them.

3.12.1 LADCs Using an Analog Logarithmic Converter

When the input and output quantities are currents or voltages an analog logarithmic converter can be obtained by means of a *logarithmic amplifier*. More generally, the analog information can be any quantity such as a frequency, a time interval etc. In this case a transducer is used appropriate to the nature of the analog signal and which obeys a logarithmic law. This type of LADC is relatively easy to make but its accuracy is limited by the use of analog techniques in the manufacture of the logarithmic converter.

3.12.1.1 Logarithmic amplifiers

The method most widely used to produce a logarithmic converter is to use a logarithmic amplifier. Considering the characteristics of the logarithmic function it is impossible to find a physical device which obeys a logarithmic law completely. Its behaviour will only be logarithmic over a limited range of the input signal which defines the *dynamic range* of the input and which is often expressed in decades.

Some logarithmic circuits can take into account the sign of the voltage to be converted and deliver a signal which can be expressed by:

$$V_S = K \ \text{sign}(V_e) \log \frac{|V_e|}{V_R}$$

K and V_R are two constants which depend upon the circuit used. In order to implement these circuits, components obeying a logarithmic law are generally used such as the p-n junction. A transistor connected as a diode can also be used as

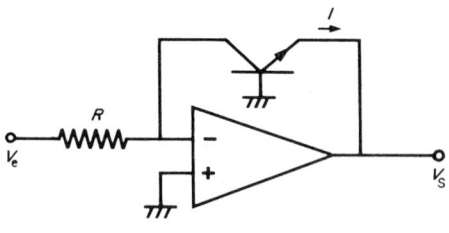

Figure 99

the feedback loop of an operational amplifier (Fig. 99). In a junction diode the current, apart from the exponential term, contains other terms due to surface losses or due to the recombination in the depletion region. In the suggested circuit these terms do not affect the output voltage V_S because the transistor voltage V_{CB}

is zero. The current is then given by:

$$I = I_{\mathrm{S}}(e^{V_{\mathrm{S}}/V_T} - 1)$$

where I_{S} and V_{T} are two constants dependent upon the transistor used.

The logarithmic law is no longer obeyed for low current values when I becomes comparable to I_{S} and for large values because of the appreciable voltage drop in the resistance of the substrate. The dynamic range of such a system may however reach nine decades.

3.12.1.2 Linear approximation by parts

A second solution consists in making a *linear approximation* by parts of the logarithmic function using for this purpose a number of function generators consisting basically of diodes and resistors. Usually each network approximates the logarithmic function over a decade and the required numbers are connected in cascade. By increasing the number of segments the discrepancy between the approximation made and the logarithmic function is reduced; for example the relative discrepancy is 8% for $n = 2$, 3.8% for $n = 3$, etc. In the compression of analog information the number of segments that must be used to approximate the logarithmic function is not usually very high. When using diode–resistor networks, changes in temperature offset the position of the break points of the characteristic obtained but do not affect the slope of the various segments. The temperature changes can be compensated by using the usual techniques.

3.12.1.3 Discharge of an RC circuit

The *discharge of an RC circuit* can also be used to obtain a logarithmic converter. This principle is of interest when the ADC which is connected after it is a voltage–frequency converter.

To begin with a capacitor C is charged by the voltage V_x to be converted and then discharged through a resistor R. The voltage across the capacitor becomes equal to a reference voltage V_{ref} after a period of time T given by:

$$T = RC \log(V_x/V_{\mathrm{ref}})$$

Time T is therefore directly related to the logarithm of V_x. It can be measured and binary encoded in the usual way by counting pulses during a given period. The manufacture of this type of converter is fairly straightforward but its accuracy is limited by several factors, among which can be mentioned the stability of the resistor, the capacitor and the reference voltage V_{ref} and also the usual factors affecting a linear ADC, the comparator, clock etc.

3.12.2 LADCs Using a Digital Logarithmic Converter

In this category of LADCs a linear ADC is followed by a *digital logarithmic converter* and the logarithmic conversion is carried out on data already quantized. All the problems related to accuracy are therefore transferred onto the ADC which has to cope with large variations of the input signal. Its resolution and accuracy are therefore limited to the number of decades corresponding to the dynamic range required for the system.

Usually a *floating point representation* is used in preference to a true logarithmic representation. In this notation f is the number of bits used for the fractional part or mantissa F, and e the number of bits used for the exponent E. A number N can be expressed as a function of f, e, F and E in several ways. In the *true floating point representation* (FP1) the expression is:

$$N = F2^E$$

In the *modified representation* (FP2) the first non zero significant bit is eliminated from the mantissa and N is expressed as:

$$N = (F + 2^f)2^{E-1}$$

The quantization error due to the compression, which can be expressed in statistical terms, can be significantly reduced if a half-interval of quantization is added to the expressions. Table 8 gives the maximum possible value of N, the number of bits b required, and the expression for the statistical error in the case of the two representations defined above.

Table 8. The dependence of three properties of LADCs on the mode of representation

Mode of representation	N_{max}	b	Statistical error
FP1	$2^{(2^e+f-1)} - 2^{(2^e-1)}$	$2^e + f - 1$	$(2\sqrt{3}\,F)^{-1}$
FP2	$2^{(2^e+f-1)} - 2^{(2^e-2)}$	$2^e + f - 1$	$(2\sqrt{3}(F + 2^f))^{-1}$

3.12.2.1 Sequential floating point converters

The simplest and by far the most used method for compressing digital information in the floating point representation consists in using sequential circuits, and in particular an arrangement of shift registers and binary counters.

For a number of n bits, compression will consist in eliminating all the most significant bits that are zero and retaining the first f bits that follow, which will represent the mantissa. Its most significant bit will always therefore be 1 and the exponent will indicate its position in the original word.

Figure 100 shows the block diagram of the principle of such a device. The content of the shift register is shifted to the right until the first non zero bit of the original word reaches the last location of the register. The mantissa is then encoded in binary in the shift register and the exponent can be obtained by counting the number of pulses used for shifting the content of the register and taking the complement of this number. To avoid the counter overflowing, an arrangement detects when E exceeds it maximum value E_{max}.

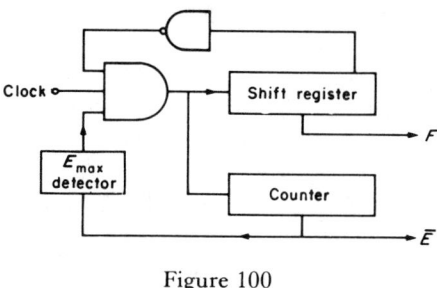

Figure 100

In this system the first bit of the mantissa is always equal to 1 since its detection stops the shifting process. It can therefore be eliminated and this leads to the FP2 representation. In this case the mantissa is represented by $f + 1$ bits. In the FP1 representation the number of components needed is about $f + 2^e + e$ flip-flops and $4 + e$ gates.

3.12.2.2 Combining floating point converters

The most general compressed form of a binary number can be obtained by using a *logic matrix*, consisting of a network of switches used for carrying out combinations. Thus b_0 boolean functions can be obtained from n input variables. In order to specify the relationship between input and output numbers one can proceed as follows. First the 2^n possible input numbers are grouped into 2^{b_0} blocks each one of them consisting of a sequence of consecutive input numbers. Then a one to one correspondence is established between each one of the 2^{b_0} input blocks and the 2^{b_0} output numbers.

This method is flexible but its main drawback is its complexity. Each output line depends on all the input lines so that if b_0 and n are large the complexity of the matrix increases to such an extent that the method is unworkable. It is also possible to use ROMs organized in words addressable by the input lines, the word content being presented in parallel on the output lines.

3.12.2.3 Incremental logarithmic converter

In an incremental converter the content of a counter increases at each clock pulse and controls the input of a DAC, which supplies a signal, known as a digital ramp, which is used as a reference voltage for comparison with the unknown signal V_x. If a linear analog to digital conversion is required then a linear DAC must be used whereas a logarithmic conversion requires the use of an *exponential DAC*.

The problem that the use of a non-linear DAC will cause can be avoided if two counters operating in parallel are used. Thus an incremental logarithmic converter is obtained (Fig. 101). A standard binary counter controls the inputs of the DAC while a logarithmic counter ensures the transformation linear to logarithmic and provides the information e and f. This system then, by simply having added a logarithmic counter, provides a linear output and a floating point representation.

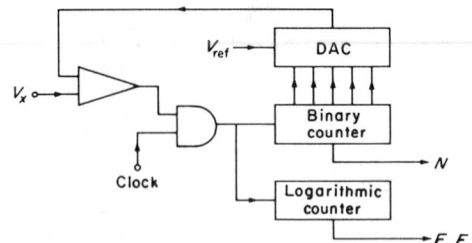

Figure 101

3.12.3 LADCs Using Non-linear DACs

One of the most used families of ADCs is the *successive approximation ADC*. A DAC inserted in the feedback loop enables a variable reference voltage to be generated which will be equal to the unknown voltage V_x to within a half quantum by the end of the conversion process. In order to make use of this family of converters for a logarithmic conversion, an *exponential DAC* must be available.

For an exponential DAC the output voltage is related to the input binary number by:

$$V_R = Q2^N = Q2^{E+F'}$$

in which Q is a constant, E corresponds to the whole part of N and is represented by e bits and F' is the fractional part of N represented by f bits. It is very difficult to implement a truly logarithmic DAC when the exponent corresponds to a large number of bits. However, a good approximation to an exponential law is obtained using the relationship:

$$V_R = QF'2^E = qF2^E$$

where $F = 2F'$ a whole binary number and $q = Q/2^F$.

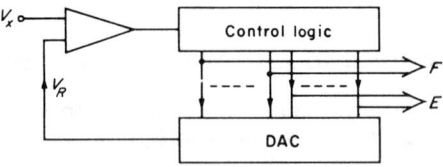

Figure 102

Figure 102 shows the block diagram of the principle of an approximation type LADC, the feedback loop consisting of an exponential floating point DAC. The conversion logic modifies the digital input message to the DAC so that the difference $V_x - V_R$ is as small as possible, then:

$$V_x = V_R = Q2^{E+F'}$$

or $E + F' = \log_2(V_x/Q)$ for a truly exponential DAC, and $V_x = qF2^E$ for the DACs actually manufactured.

Before discussing in some detail an ADC manufactured using this principle, the various solutions adopted for obtaining a floating point DAC will be examined.

3.12.3.1   Weighted resistor exponential DAC

As in the case of a linear conversion, it is possible to use weighted resistors or ladder networks and associated switches to obtain an exponential DAC. Figure 103 gives the basic schematic diagram of an exponential DAC using *weighted*

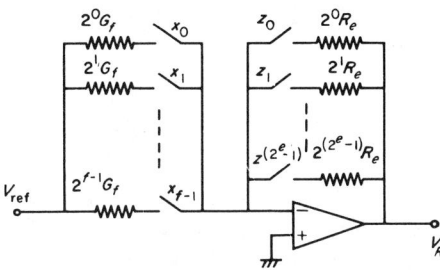

Figure 103

resistors. The floating point representation of the number $N = F2^E$ is obtained using the two relationships:

$$F = \sum_{i=0}^{f-1} 2^i x_i \qquad E = \sum_{i=0}^{2^e - 1} 2^i y_i$$

f being the number of bits used to represent the mantissa and e that used to represent the exponent. The bit of rank i of the mantissa acts directly on the corresponding switch which is closed if $x_i = 1$ and if not the switch is open. The switches corresponding to the exponent which put in circuit a feedback resistor of $2^i R_e$ $(0 \leqslant i \leqslant 2^e - 1)$ are mutually exclusive and are controlled by a logic function z_j such that:

$$z_j = 1 \quad \text{if } j = E$$

$$z_j = 0 \quad \text{if } j \neq E$$

The output voltage obtained can be expressed by:

$$V_R = V_{ref} \frac{G_f}{G_e} 2^E \sum_{i=0}^{f-1} 2^i \cdot x_i$$

$$= V_{ref} \frac{G_f}{G_e} F2^E$$

By comparing it with the previous expression of V_R the value of the quantum is obtained:

$$q = V_{ref} \frac{G_f}{G_e}$$

In order that the full-scale range of V_R be equal to V_{ref} it is necessary that:

$$V_{R\,max} = V_{ref} = V_{ref} \frac{G_f}{G_e} F_{max} 2^{E\,max}$$

or

$$\frac{G_f}{G_e} = \frac{1}{F_{max} 2^{E_{max}}} = \frac{1}{(2^f - 1)2^{(2^e - 1)}}$$

V_R can then be written:

$$V_R = V_{ref} \frac{F2^E}{(2^f - 1)2^{(2^e - 1)}}$$

As in the case of a linear DAC it is difficult to manufacture resistors of widely different values. Now the ratio between the extreme resistance values must be equal to $2^f + 2^e - 1$ and this very quickly limits the values of f and e.

3.12.3.2 Exponential DACs using ladder networks

It is also possible to obtain exponential DACs by using *ladder networks* as shown by the block diagram of Fig. 104. The network operates normally giving an output voltage V_0:

$$V_0 = V_{ref} \frac{F}{2^f}$$

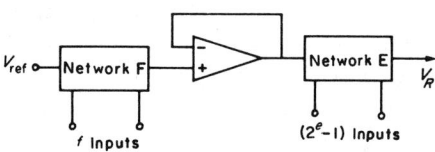

f Inputs $(2^e - 1)$ Inputs

Figure 104

According to the theory developed for the weighted resistors exponential DACs, network E is controlled by 2^e signals obtained by decoding the exponent E consisting of e bits. The analog voltage applied to the input of network E is the output voltage V_0 of network F (the amplifier has a gain of unity). The voltage V_R is given by:

$$V_R = \frac{V_{ref}}{2^f 2^{2e}} F2^E = qF2^E$$

Note that in this arrangement network E will work with input voltages V_0 which could be as low as $V_{ref}2^{-f}$. This imposes limits upon the admissible errors due to network E and associated switches. However, this is consistent with the fact that the aim is to achieve the same accuracy as that obtained with a linear converter of $(f+2^e)$ bits. This remark applies in general to LADCs for which a wide input dynamic range is required. For low voltages this necessitates a very high accuracy.

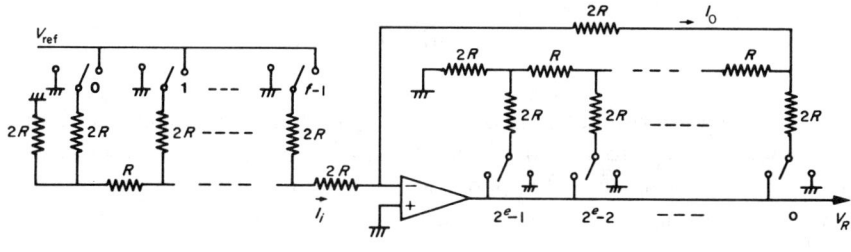

Figure 105

Figure 105 shows another way of using ladder networks in the design of an exponential DAC. Network F supplies a current I_i given by:

$$I_i = \frac{1}{2} \frac{V'_{ref}}{3R} \frac{F}{2^f}$$

Current I_0 and voltage V_R are related by:

$$I_0 = \frac{1}{2} \frac{V_R}{3R} 2^{-E}$$

Equating the expressions for I_i and I_0 gives,

$$V_R = \frac{V'_{ref}}{2^f} F2^E$$

In order that full-scale V_R be equal to V_{ref} it is necessary that:

$$V'_{ref} = \frac{V_{ref}}{2^{2e}}$$

The advantages of this twin network system lie in the possibility of integrated circuit fabrication of the resistors and associated switches and above all in the fact that only two values of resistances, R and 2R, are required. This allows the temperature drift to be minimized.

3.12.3.3 LADCs using an exponential DAC

Among the different existing families of ADCs a certain number make use of a DAC in the feedback loop. Using the same principle LADCs are obtained provided an exponential DAC is used. Only two examples will be given.

The first example is that of the *counter ramp converter* (also called *digital ramp converter*) (Fig. 106). The content of the binary counter is divided into two parts;

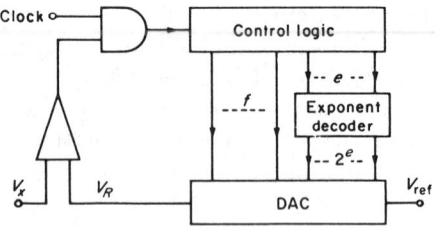

Figure 106

one part corresponds to the mantissa and the other to the exponent. The e bits of the exponent are sent to a decoding arrangement having 2^e outputs only one of which is in state 1 (the one corresponding to the value E) and which controls one of the two resistor networks of the DAC. The clock pulses are sent to the counter until the condition

$$V = V_D = qF?^E$$

is satisfied. In this case a DAC with $(f + 2^e)$ inputs must be used. This system is very simple but as is the case for its linear equivalent a significant time is needed to carry out one conversion.

Again the most useful system is the *successive approximation system*. The conversion logic used is similar to the conversion logic used for the linear case. At the first clock pulse the bit of greatest weight of the fractional part (mantissa) is set at 1 and the unknown voltage V_x is compared to $V_R = 2^{f-1}$. If the comparator detects that $V_x > V_R$ the value of the exponent E is modified by increasing it starting from zero until it satisfies the condition:

$$V_x < q 2^{(f-1)} 2^E$$

The value of E is then maintained constant and the value of the various bits of F determined by successive tries as in the case of a linear system. If during the search for the value of E the maximum value $2^e - 1$ is reached and V_x is still greater than V_R, E is then held at its E_{max} value and the value of the various bits of F determined as above.

If at the start of the conversion it is noticed that $V_x < V_R = q 2^{f-1}$ then the bit of greatest weight is reset to zero and the second is tried, while keeping E equal to zero. The procedure is repeated until $V_x > qF$ is obtained. Then the value of the exponent is found and the conversion completed as above. This system, like the linear converters is of great interest since it combines a medium level of complexity with high speed and very good accuracy.

4 DIGITAL–SYNCHRO AND SYNCHRO–DIGITAL CONVERTERS

4.1 ANGULAR SENSORS

Before studying these different types of converters it is useful to describe various sensors enabling the angular position of an axis or shaft to be known.[18,52]

4.1.1 Disc Encoders

A *disc encoder* is a sensor which converts a quantity related to the rotation of a shaft directly into a number. This encoder gives access to two variables: the angular position and the angular speed. Two types of disc encoders can then be distinguished.

Incremental encoders which generate one pulse each time an angle θ has changed by $360/2^n$ in either direction. For these types of encoders n is usually between 8 and 12.

Absolute encoders which generate unique digital words representing the angle θ using n bits. In this case a reference position must be specified.

4.1.1.1 Incremental encoders

Figure 107 gives a drawing of a two track incremental encoder. A rotating disc has two rings divided into 2^n equal sectors. The angle α between sectors is therefore $360/2^n$ degrees or $21\,600/2^n$ minutes of arc. These two rings are displaced from each other by a quarter of a sector. Each sector is divided into two areas, one of which is conducting and the other is not. One can choose to sense light or electricity and, depending on the choice, contacts or opto-electronic coupling devices are used to read out the information.

Each incremental rotation from one sector to the next is counted as a pulse. Hence the signal obtained has a rate of change N times greater than that of the disc. Therefore, by using a counter, the number of shaft rotations can be counted

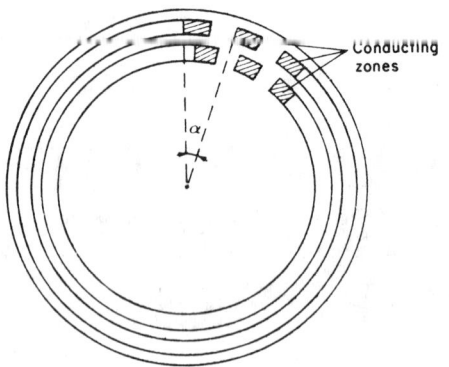

Figure 107

or its speed of rotation measured. If the direction of rotation is required then an up-down counter is sufficient together with the read out of the second track while the analysis of the simultaneous read out of both tracks yields the direction of rotation of the shaft.

4.1.1.2   Absolute encoders

When it is required to know the absolute value of an angle, an n-track encoder is used. The track of rank i has 2^i zones alternately conducting and insulating. The outer track has 2^n zones and corresponds to the LSB whereas the innermost track corresponds to the MSB and has only two zones. The resolution that can therefore be obtained is $360°/2^n$. Figure 108 shows a four bit encoder. To obtain a digital read out of angle θ, n electrical or optical sensors per track must be available. When the natural binary code is used this system has the drawback that when one passes from a number N to $N+1$ or $N-1$ it may happen that several bits change state simultaneously. This transition can be fairly slow with an

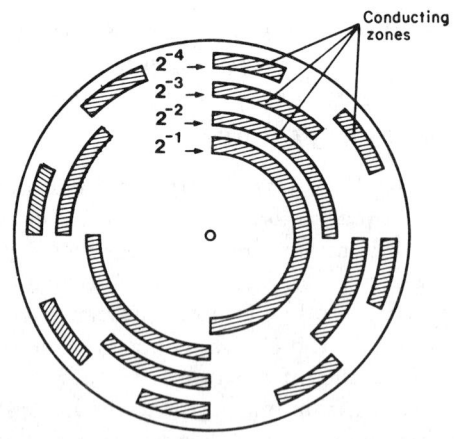

Figure 108

electromechanical arrangement which furthermore could stop at the instant of time the transition is taking place. Hence it is necessary to remove the ambiguity of the reading. This can be achieved by using the Gray code or by doubling the number of sensors reading the information, these being positioned according to a precise pattern (U or V scan).

4.1.2 Synchros and Resolvers

Another method of transmitting the position of a shaft is to use a synchro device. The angular data is then transmitted as alternating voltages and this information is accepted at the receiving end by a synchro receiver. Resolvers and synchros are grouped in this category.

A synchro transmitter, or synchro for short, consists of a single phase rotor fed from an AC reference voltage (usually at 400 Hz) and a stator having three windings displaced from each other by 120°. (Fig. 109). The voltages induced in

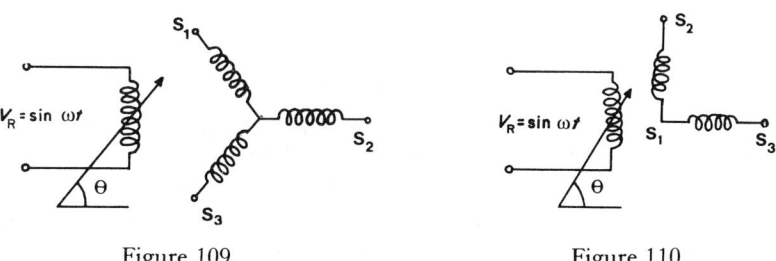

Figure 109   Figure 110

these three windings are sinusoidal and their amplitude depends upon the angle θ of the rotor position:

$$V_{S1} - V_{S2} = K_1 \sin \theta \sin(\omega t + \alpha_1)$$

$$V_{S2} - V_{S3} = K_2 \sin(\theta + 120) \sin(\omega t + \alpha_2)$$

$$V_{S3} - V_{S1} = K_3 \sin(\theta - 120) \sin(\omega t + \alpha_3)$$

For an ideal synchro the constants K_1, K_2, K_3 are equal and the angles $\alpha_1, \alpha_2, \alpha_3$ are zero. In practice various errors result in different constants K and non-zero angles α. For a resolver the stator has two windings at 90° (Fig. 110). The rotor induces in these windings two voltages proportional to the sine and cosine of the angle θ thus:

$$V_{S1} - V_{S2} = K_1 \sin \theta \sin(\omega t + \alpha_1)$$

$$V_{S1} - V_{S3} = K_2 \cos \theta \sin(\omega t + \alpha_2)$$

Constants K_1 and K_2 are equal for an ideal resolver as are the angles α_1 and α_2. Thus the information provided by a synchro or a resolver is in the form of AC signals of constant amplitude for a given angle θ.

Certain ratios can be established which are independent of the carrier signal and depend only upon the angle θ. For a resolver:

$$\frac{V_{S1} - V_{S2}}{V_{S1} - V_{S3}} = \tan \theta$$

For a synchro:

$$\frac{V_{S1} - V_{S2}}{V_{S2} - V_{S3}} = \frac{\sin \theta}{\sin(\theta + 120)}$$

$$\frac{V_{S2} - V_{S3}}{V_{S3} - V_{S1}} = \frac{\sin(\theta + 120)}{\sin(\theta - 120)}$$

$$\frac{V_{S3} - V_{S1}}{V_{S1} - V_{S2}} = \frac{\sin(\theta - 120)}{\sin \theta}$$

When θ varies these ratios also vary, but will yield the value of θ whatever the speed or acceleration of the shaft. In particular, in the case of a resolver, θ is always given by:

$$\theta = \tan^{-1}\left(\frac{V_{S1} - V_{S2}}{V_{S1} - V_{S3}}\right)$$

In fact when the angle θ varies, parasitic voltages are generated in the stator proportional to the rate of change of θ, are added to the voltages defined previously and distort the measured value of θ.

There are circuits available which will transform the signals sent by a resolver in order to apply them to a synchro and vice versa. One such passive circuit is

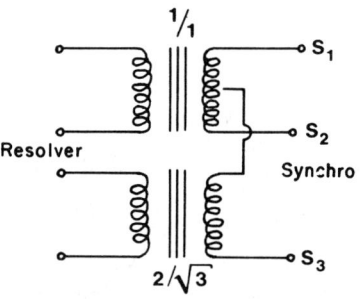

Figure 111

given on Fig. 111 and is known as a *Scott connected transformer*. It uses two transformers having ratios of $1:1$ and $2:\sqrt{3}$. The voltage equations are given by:

$$V_1 - V_2 = E \sin \theta \sin \omega t$$

$$V_3 - V_1 = -E\left[\frac{1}{2} \sin \theta + \frac{\sqrt{3}}{2} \cos \theta\right] \sin \omega t$$

$$= E \sin(\theta - 120) \sin \omega t$$

$$V_2 - V_3 = E\left[-\frac{1}{2}\sin\theta + \frac{\sqrt{3}}{2}\cos\theta \right]\sin\omega t$$

$$= E\sin(\theta + 120)\sin\omega t$$

Figure 112 shows an electronic circuit, called an electronic Scott circuit, which transforms three phase information (synchro) into two phase information (resolver). For the reverse transformation circuit, phase S_1 is taken as the origin and connected to the reference point.

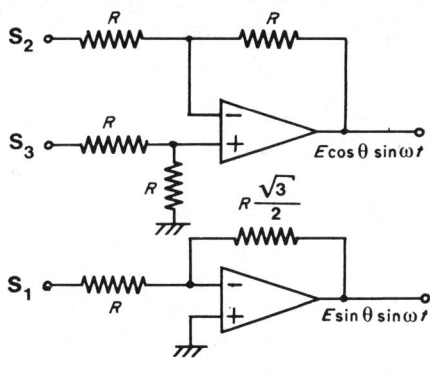

Figure 112

4.1.3 Comparison

It is possible to compare the characteristics of these different encoders. The most reliable are the synchros and resolvers which are not sensitive to ageing or to temperature, next the optical encoders which are sensitive to temperature and to the possible presence of electromagnetic fields and finally the potentiometers which have a limited period of life due to contact wear.

In order to compare costs it is necessary to include the associated electronic equipment and to choose the same resolution. Thus, listing the devices in decreasing order of costs gives: optical encoders, synchros and resolvers and then potentiometers. The performance that can be obtained as regards resolution, accuracy and dynamic response are comparable for the optical encoders and synchros or resolvers and slightly lower for the potentiometers.

4.2 DIGITAL–AC CONVERTERS

It is often necessary to have AC signals available when the control of electromechanical devices is required. It is therefore advantageous to generate sinusoidal signals whose amplitude is proportional to the number corresponding to the digital control signal. The operation of digital–AC converters (DAICs)[2,10,F2,F3] differs little from the operation of most of the DACs already discussed. A case of special interest is that of digital–synchro converters for the direct control of synchros or resolvers. Many DAICs are parallel decoders because the control signals of electromechanical systems need to be continually available and it is very difficult to store AC signals in a memory. Digital AC converters need not be very fast since electromechanical systems are usually slow. These systems require appreciable power for the control circuits hence the DAICs are often followed by power amplification states. DAICs require an alternating reference voltage which is usually available in electromechanical systems and its frequency is generally chosen to be 400 Hz.

4.2.1   DAICs Characteristics

The characteristic parameters of DAICs are identical to the characteristic parameters of the DACs previously described. In this case the reference quantity is equal to 360°. The resolution corresponds also to the number of bits n that the converter accepts and the value of the quantum is $360°/2^n$.

The amplitude of the output alternating voltage of the DAIC will be proportional to the value of θ introduced at the input of the converter in a digital form. The conversion or settling time is the time required to obtain the alternating output signal corresponding to the digital input signal within the specified accuracy. The conversion time takes into account the time needed for the transient condition to die down. The problem of 'glitches' does not arise with DAICs because the time constants of electromechanical systems are large and often act as low-pass filters.

The errors which limit the accuracy of DAICs are the gain, offset, non-linearity and non-monotonic errors, the noise generated by the converter and the errors due to the function generator whose role will be defined further on. For digital–synchro converters the load connected to the output of the converter may introduce further errors such as an increase of the non-linearity, a worsening of the rate of change of the output, lower accuracy because of poorer gain etc.

4.2.2   DAICs which use a Parallel DAC

The first category of DAICs is made up of multiplier DACs, that is DACs which accept a variable reference voltage. Most of the DACs examined so far can be used as DAICs. However, for this application it is sometimes necessary to modify the switches used because the amplitude of the reference voltage may be appreciable (of the order of 100 V) and this necessitates the use of special switches. Weighted resistor DACs or ladder type DACs can be used for DAICs provided the precaution is taken to use series–parallel switches. These switches are often FETs and the resistances presented by these transistors when they are conducting must be compensated for. This is done by inserting an FET in the conducting state into the feedback loop of the amplifier at the converter output.

Inverted ladder DACs need no modification other than that of the reference voltage and this is an added advantage of this group of converters.

The use of a DAC as a DAIC usually results in a greater complexity than for normal DAC operation, assuming of course that the accuracy is kept the same. As the reference voltage varies with time it is necessary to use switches whose accuracy does not depend upon the magnitude of the voltage (or current) to be switched. The offset voltages do not introduce any error because the output voltage is often made available via a transformer and moreover the accuracy can be limited because of the use of a power amplifier.

4.2.3   DAICs for DC–AC Conversion

When the output of a DAC supplying DC is connected to a DC–AC converter the arrangement thus formed constitutes a second category of DAIC. This technique can be used with all the DACs already described but is only really of interest if a DAC is already available in the circuit under consideration.

A DC–AC converter is a device which receives a DC signal (voltage or current) supplied by the DAC and generates an AC signal, or modulates its amplitude so that it is proportional to the DC input signal. Many magnetic converters carry out this function but they are bulky, heavy and costly.

Another very simple method of obtaining a sinusoidal signal of the required amplitude consists in modulating the DC output signal of the DAC by a periodic signal. This can be done by means of a multiplier or a chopper. Figure 113 gives

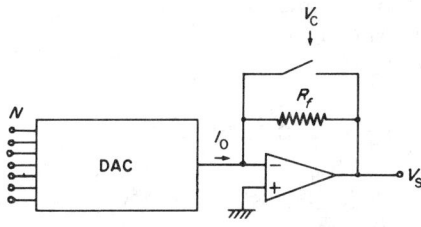

Figure 113

the basic schematic diagram of such a device. The feedback resistor of the output amplifier of the DAC used is shunted by a switch, which is controlled by a square wave form whose amplitude is not important, but whose frequency must be equal to the frequency required for the output signal V_x. When the switch is open the output voltage obtained is:

$$V_S = -R_f I_0$$

When the control signal is applied the output gives square signals of amplitude V_x and of a frequency equal to the frequency of the control signal (Fig. 114). All that

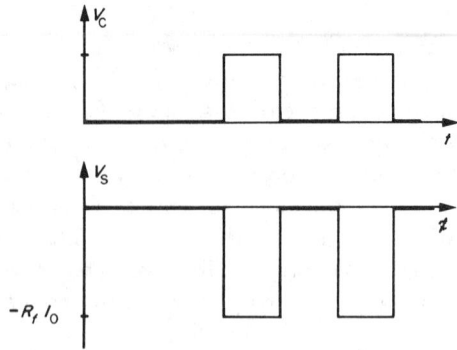

Figure 114

need be done to obtain a sinusoidal signal is to connect a low-pass filter (an RC circuit for example) after the device.

When the input message N varies, the current I_0 also varies as does the amplitude of voltage V_S thus resulting in a change of the sinusoidal amplitude. Most electromechanical systems have large time constants and it is then possible to use the square wave signals directly to control the system which carries out its own filtering operation.

If a DC voltage V_S and a sinusoidal voltage $E \sin \omega t$ are both applied to a multiplier the output signal will be of the form

$$K V_S \sin \omega t$$

If a multiplier is used there is no point in filtering, but the amplitude E of the sinusoidal signal must be maintained fairly constant otherwise errors will be present in the output of the DAIC.

4.2.4 DAICs Using a Transformer

This method of conversion gives a better accuracy (up to 16 bits) but the cost and bulk of the equipment is greater. In this converter a transformer is used whose primary is fed by the reference voltage (usually at 400 Hz) and whose secondary has as many coil windings as there are bits in the word to be converted (Fig. 115). The number of turns of these different coils form a geometric progression of common ratio two. The voltages induced in the various secondary coils, ignoring the coefficient of proportionality, are then fractions of the reference voltage:

$$\frac{V_{ref}}{2}, \quad \frac{V_{ref}}{4}, \quad \frac{V_{ref}}{8}, \cdots \frac{V_{ref}}{2^n}$$

The coils are interconnected as required to give an output voltage of amplitude proportional to the word N to be converted, which appears between A and B. In

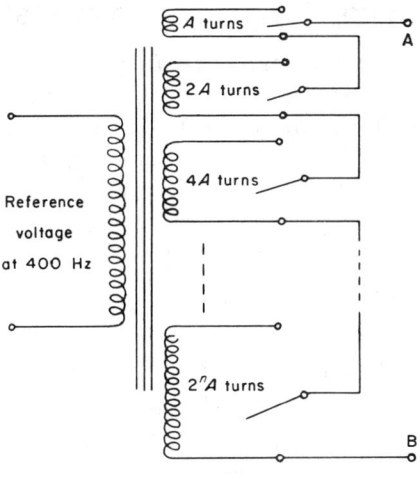

Figure 115

order to interconnect the coils, series–parallel switches are used which have to withstand high voltages and switch large currents. Transformer coupled transistor switches appear to be the most advantageous for this use since, in particular, they provide good isolation between the transistor switches and the control circuit which is achieved by two transformers. Using a transformer with n coils gives great accuracy since, in principle, it only depends upon the number of turns of each coil, a number known accurately. This type of converter is very slightly sensitive to changes in the ambient environment (temperature etc.). Its overall accuracy depends upon the resistances presented by the switches and the series resistance of the various interconnected windings. The linearity depends only upon the difference between the resistances presented by switches in the series and parallel branches and not their absolute value. An accuracy of 10^{-3} is easily obtained with this method.

4.2.5   Digital–Synchro Converters

Digital–synchro converters (DSCs) constitute a special category of DAICs. The digital information N that is to be converted corresponds to the value of an angle, θ, and it is required to obtain two sinusoidal voltages, proportional to $\sin \theta$ and $\cos \theta$ respectively. By means of these two voltages it will then be possible to control a resolver or a synchro using if need be a Scott connected transformer.

Although these converters provide only two voltages (which allow them to control a resolver) they carry the name of digital–synchro converters. They make use of techniques described previously and are differentiated by the manner in which the information θ is transformed into the two quantities $\sin \theta$ and $\cos \theta$. Depending upon the circumstances, these two quantities are used in analog or digital form.

4.2.5.1   Sin θ and cos θ available in digital form

It will be assumed first that sin θ and cos θ are available in digital form and this leads us to a first family of DSCs. These two pieces of information are used for controlling two DACs which can be two DACs providing a DC output or two multiplier DACs. Figure 116 shows the block diagram of a DSC using two

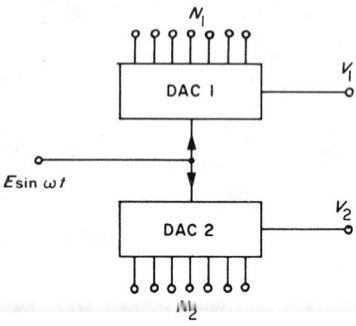

Figure 116

multiplier DACs which are fed with the same reference voltage, $E \sin \omega t$, and two digital inputs N_1 and N_2 corresponding to sin θ and to cos θ. The two analog voltages available at the outputs of the two DACs are:

$$V_1 = K_1 E \sin \theta \sin \omega t$$

$$V_2 = K_2 E \cos \theta \sin \omega t$$

In order that no error occurs the two constants K_1 and K_2 introduced by the two DACs must be identical, because the resolver determines the angle θ it must display from the ratio:

$$\frac{V_1}{V_2} = \frac{K_1}{K_2} \tan \theta$$

The value of amplitude E has no effect on this ratio which eliminates any spurious variations of E.

A second solution consists in using two DACs supplying DC voltages (Fig. 117). The two numbers N_1 and N_2 proportional to sin θ and cos θ are available at the inputs of the two DACs and thus two DC voltages V_1 and V_2 are obtained:

$$V_1 = V_{\text{ref}} \sin \theta$$

$$V_2 = V_{\text{ref}} \cos \theta$$

These two voltages are then applied to the inputs of two modulators which also receive the same sinusoidal voltage $E \sin \omega t$. The two output voltages are given by:

$$V_{S1} = k_1 V_{\text{ref}} \sin \theta \sin \omega t$$

$$= K_1 \sin \theta \sin \omega t$$

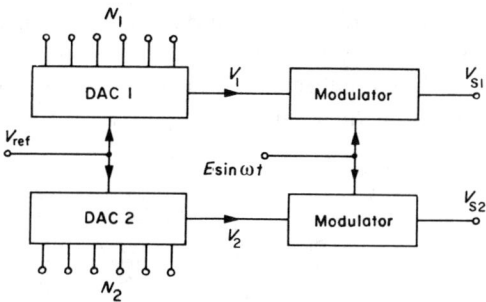

Figure 117

$$V_{S2} = k_2 V_{ref} E \cos \theta \sin \omega t$$

$$= K_2 \cos \theta \sin \omega t$$

As before this connection eliminates the fluctuations of the reference voltages V_{ref} and $E \sin \omega t$ which are common to the two channels.

4.2.5.2 Sin θ and cos θ available in analog form

When the pieces of information sin θ and cos θ are available in analog form it is no longer necessary to use DACs. The information can then be applied directly to the inputs of two modulators fed from the same reference voltage $E \sin \omega t$ as in the case of Fig. 117.

4.2.5.3 Obtaining sin θ and cos θ

The above circuit assumed that the two pieces of information sin θ and cos θ were available in digital or analog form, obtained from the values of an angle θ available in digital form. It is now necessary to examine how these two pieces of information can be obtained. The systems providing this operation are called function generators. The block diagram of Fig. 118 can be used when it is required to obtain analog quantities proportional to sin θ and cos θ. The digital information is first transformed into a DC voltage V_1 proportional to θ by means of a DAC. This voltage is then applied to the input of two analog function generators, which

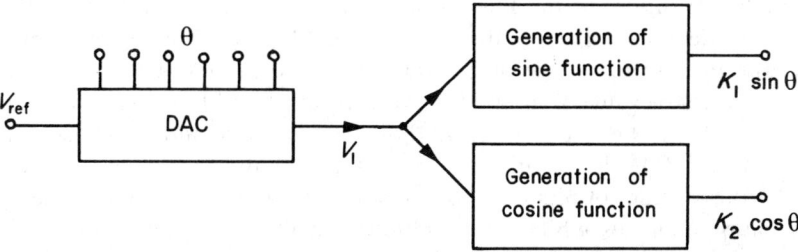

Figure 118

deliver two voltages proportional to $\sin\theta$ and $\cos\theta$. The function generators are made up of diodes, resistors and voltage sources as shown on the diagram of Fig. 119. When the voltage V_E increases diodes D_1, D_2, etc. begin to conduct and the introduction into the circuit of resistors R_1, R_2, etc., modifies the slope of the

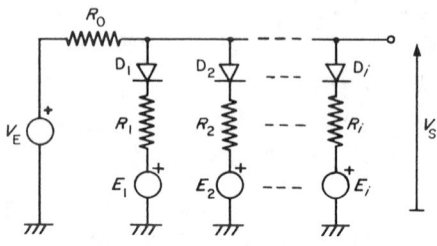

Figure 119

characteristic $V_S = f(V_E)$. The values of the resistors and of the voltage sources are chosen so that the output voltage V_S follows the law required. As an example Table 9 gives the values that the various components must have in order that voltage V_S varies according to a sinusoidal law with an accuracy of the order of 1%. The values are expressed in terms of R_0 and the maximum value $V_{E\max}$ of V_E.

Table 9. The values of various components which allow V_S to vary according to a sinusoidal law within an accuracy of one per cent

Section number	Value at the break point of $V_E/V_{E\max}$	Value of $E_i/V_{E\max}$	Value of R_1/R_0
1	0.28	0.27	32.2
2	0.44	0.41	6.25
3	0.61	0.52	3.45
4	0.78	0.60	1.42
5	0.88	0.63	0.355

Some of the DSCs described above require that the information $\sin\theta$ and $\cos\theta$ be available in digital form. Therefore a device that generates these quantities from the value of an angle θ in digital form must be used. These digital function generators usually make use of ROMs. The word N fed at the input (which corresponds to the value of the angle θ) constitutes the address of a location in the memory in which the information $\sin\theta$ or $\cos\theta$ is stored and which is then read out. In a way a ROM is a digital conversion table having 2×2^n memory locations (n being the number of bits used for encoding the angle θ). The word length of each location is usually n bits, for it is required to obtain the values $\sin\theta$ and $\cos\theta$ to the same accuracy as that of angle θ. The signals from the memory can be applied to DACs which supply DC voltages, or to DACs such as those which have

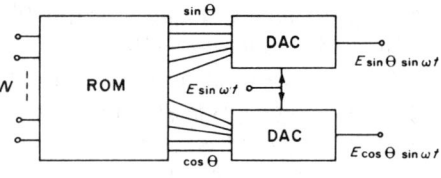

Figure 120

been described above. Figure 120 shows the block diagram for a DSC with memory.

When the number of bits, n, used for representing the angle θ becomes large, the size of the memory required to store all the values of $\sin \theta$ and $\cos \theta$ can become excessive. It can be reduced by using a *quadrant selector*. The two most significant bits of the word N represent the quadrant (or quarter of circle) in which angle θ lies. The remaining $(n-2)$ bits correspond then to an angle less than 90° which shall be designated by α. It is possible to express in a simple way $\sin \theta$ and $\cos \theta$ as a function of $\sin \alpha$ and $\cos \alpha$. The block diagram of converter with a quadrant selector is shown in Fig. 121. The two bits of greatest weight

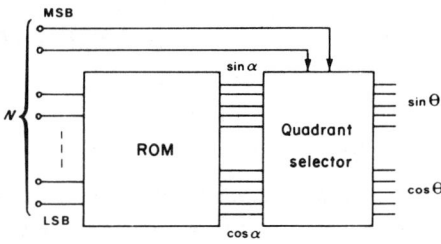

Figure 121

are used to determine the quadrant in which the angle θ lies. Using the $(n-2)$ bits corresponding to an angle α a ROM supplies the information $\sin \alpha$ and $\cos \alpha$ in digital form. The quadrant selector receives the information $\sin \alpha$ and $\cos \alpha$ and the two bits of greatest weight, and then generates the information $\sin \theta$ and $\cos \theta$ using the results of Table 10. The two signals thus obtained are sent to the input

Table 10. A conversion table to generate the value of $\sin \theta$ and $\cos \theta$ from the value of α

Quadrant	Value of bits of greatest weight		$\sin \theta$	$\cos \theta$
1. $0 \leqslant \theta < \pi/2$	0	0	$\sin \alpha$	$\cos \alpha$
2. $\pi/2 \leqslant \theta < \pi$	0	1	$\cos \alpha$	$-\sin \alpha$
3. $\pi \leqslant \theta < 3\pi/2$	1	0	$-\sin \alpha$	$-\cos \alpha$
4. $3\pi/2 \leqslant \theta < \pi$	1	1	$-\cos \alpha$	$\sin \alpha$

of two DAICs as before. The advantage of using the quadrant selector is that the size of the memory is reduced by a factor of $2 \times 2^2 = 8$.

The block diagram of Fig. 122 can also be used. In this case the quadrant selector delivers the signals sin θ and cos θ which are applied to the input of two DAICs. The AC voltages which are used as reference voltages for these converters are chosen to be in phase or out of phase depending upon the signs of sin θ and cos θ. To do this the DAIC is fed through a transformer whose secondary winding has a mid-point connected to earth and one or other winding end is chosen depending upon the sign of sin θ and cos θ.

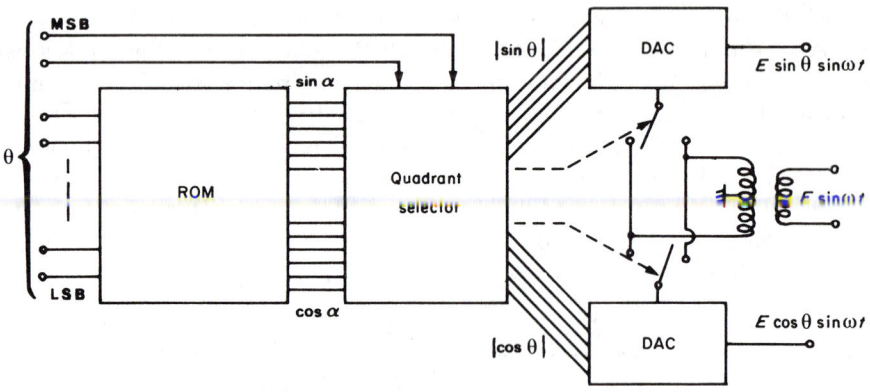

Figure 122

Instead of a quadrant selector an *octant selector* can be used. The first three bits are then used for determining the octant (or $\frac{1}{8}$ part of circle) in which the angle θ lies. The remaining $(n - 3)$ bits correspond to an angle α less than 45°. The values sin α and cos α are generated using a ROM whose size will be further reduced and the octant selector will give the two pieces of information sin θ and cos θ from the values of sin α and cos α and the information contained in the first three bits.

4.3 AC–DIGITAL CONVERSION

4.3.1 General

Electromechanical sensors are used in many control or command systems because they are economical and accurate. With the ever increasing use of computers in these systems it becomes necessary to be able to convert easily the information provided by these sensors into digital information. The circuits enabling alternating signals to be converted[2,18,52,53] can be grouped into three categories:

converters transforming a single phase alternating signal into digital information,

converters receiving signals from a synchro or a resolver and producing the
information $\sin \theta$ and $\cos \theta$ in digital form,
synchro–digital converters producing the information θ in digital form.

The signals from sensors are all amplitude modulated. Thus the information to be
digitized is contained in the amplitude of the signal to be processed. The
expressions for the various signals which may have to be processed are given
below:

single phase signal

$$V_{(t)} = E \sin(\omega t + \phi)$$

two-phase signal

$$V_1 = K_1 E \sin \theta \sin \omega t$$

$$V_2 = K_2 E \cos \theta \sin \omega t$$

three-phase signal

$$V_1 = K_1 E \sin \theta \sin \omega t$$

$$V_2 = K_2 E \sin(\theta - 120) \sin \omega t$$

$$V_3 = K_3 E \sin(\theta + 120) \sin \omega t$$

For an ideal sensor the coefficients K are equal, the amplitude differences being
only due to the value of the angle θ. In order to pass from a 2-phase to a 3-phase
signal and vice versa all that is required is a Scott connected transformer. It must
be remembered that a synchro or a resolver introduces only a spatial phase change
(and not a time phase change) with respect to the reference signal $\sin \omega t$. The
possible phase changes of the signals produced must be considered to be parasitic.
Moreover in most cases it is not the absolute value of the amplitude that matters
but simply its value with respect to a reference amplitude. Thus the sensors and
converters that follow them will not be sensitive to the changes of such quantities
as the frequency of the reference signal, its amplitude, etc.

4.3.2 AC–DC Conversion

Many AC digital conversion systems first convert the AC signals into DC voltages
before transforming them into digital quantities. Therefore this operation, called
demodulation, must be examined first.

   If the DC signal obtained is a function of the phase relationship existing
between the phase of the signal to be demodulated and that of the reference signal
then it is referred to as a phase sensitive demodulator. In general the signals to be
demodulated are in phase or out of phase with the reference signal, and a phase
shift with respect to these two conditions is then considered to be an error.
Demodulation is sometimes called crest or peak detection because a voltage is
obtained whose value is proportional to the amplitude and therefore to the crest or
peak value of the AC signal.

The relationship between input and output signals of a demodulator can be written: $V_{DC} = K V_{AC} \cos \phi$ where ϕ is the phase difference between the signal to be demodulated and the reference signal, V_{AC} is the amplitude of the AC signal and K is a constant depending upon the circuit used.

Usually a demodulator is made up of an inverter, a switch, a zero detector and a low-pass filter as shown in Fig. 123. When the polarity of the reference voltage

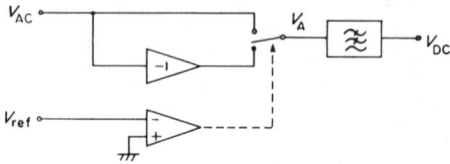

Figure 123

changes the switch operates so that the voltage V_A applied to the input of the low-pass filter always keeps the same sign (+ or − according to whether the signals V_{AC} and V_{ref} are in phase or out of phase). Many circuit arrangements will carry out these operations. A first solution consists in using a transformer whose secondary winding has a mid-point, the voltage V_{AC} being applied to the primary. Each half of the secondary winding is connected to the input of the filter through a switch and both switches are controlled by the reference signal in such a way that they are mutually exclusive and in this way very good accuracy is obtained. The main drawback of the system is in the size and weight of the transformer particularly when dealing with low frequency signals.

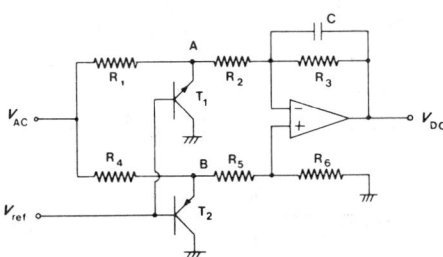

Figure 124

The second circuit arrangement shown in Fig. 124 does not use a transformer, which makes the device lighter, but it is less accurate and more sensitive to noise than the previous circuit. It can be looked upon as a differential amplifier. When the reference voltage is positive transistor T_1 is saturated and point A is at zero potential, the amplifier then behaves as a follower circuit having a gain G_1:

$$G_1 = \frac{R_6}{R_4 + R_5 + R_6} \cdot \frac{R_2 + R_3}{R_2}$$

In a similar way when the reference signal is negative, transistor T_2 is saturated and point B in its turn is at zero potential. The system behaves as an inverter circuit of gain G_2:

$$G_2 = \frac{R_3}{R_1 + R_2}$$

The impedances, as seen by the amplifier inputs, are different depending whether it is the + input or the − input. Thus

$$R- = R_4/(R_1 + R_2)$$
$$R+ = (R_4 + R_5)/R_6$$

For the output signal V_{DC} to be independent of the phase of the reference voltage and therefore of the path taken, it is necessary that:

$$G_1 = G_2 \quad \text{and} \quad R- = R+$$

which imposes some restrictions on the resistor values. Capacitor C connected across R_3 provides the filtering.

Synchronous detection can also be used in order to carry out the operation of demodulation. In this case the AC signal received is multiplied by the reference voltage and the DC component is separated from the resulting signal by a low-pass filter. In contrast to what is obtained when the systems described above are used, the recovered DC voltage V_{DC} is a function of the amplitude of the reference voltage which therefore must be stable if no error is to be introduced. For this reason synchronous detection is little used in AC digital conversion.

4.3.3 Single Phase Converters

One way of transmitting information by means of a single phase signal is to modulate the amplitude of this signal with the useful information. For example when it is required to transmit the value of an angle θ with one signal an inductive sensor is used which provides an AC signal proportional to θ. The first family of AC digital converters (AIDC) therefore includes converters processing single phase signals.

The first method used to convert a single phase signal consists in using two phase sensitive demodulators according to the block diagram of Fig. 125. The arrangement also uses an ADC which can accept an external voltage reference.

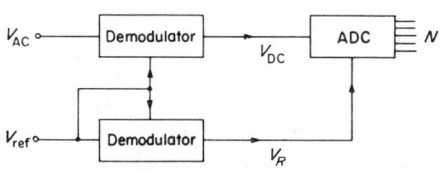

Figure 125

Both the AC signal to be converted and the reference signal are sent to both demodulators which supply two DC voltages, V_{DC} and V_R proportional to the above two signals. The amplitude corresponding to the useful signal V_{DC} is positive or negative depending whether the AC signal is in phase or out of phase with the reference signal. This voltage V_{DC} is then applied to the input of the ADC which moreover receives the voltage V_R. The digital information supplied by the ADC is then proportional to the information to be transmitted. Freedom from parasitic changes in the amplitude of the reference voltage is obtained with this arrangement, because the parasitic changes affect in the same way both voltages V_{DC} and V_R and are therefore eliminated by the ADC. The performance of this ADC corresponds to the 'sum' of the performance of the component parts and is usually limited by the performance of the demodulators.

Another method consists of sampling the AC signal to be converted as shown in Fig. 126. The peak detector generates a control signal V_C which is applied to

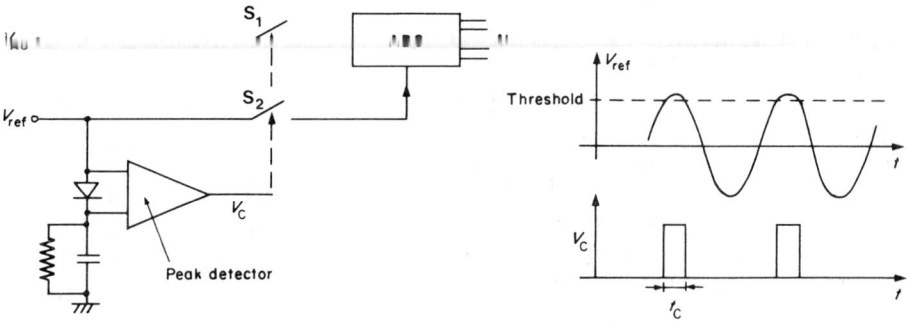

Figure 126

switches S_1 and S_2 by comparing the reference signal V_{ref} to a given threshold. In order to eliminate the changes in the reference signal the threshold is generated from that signal. Switches S_1 and S_2 apply the signals V_{AC} and V_{ref} to the ADC during the time t_C of the control signal. This time must not be too short, so that the converter has time to operate, nor too long, in order that the changes in the signals V_{AC} and V_{ref} be compatible with the accuracy required. This method also gives good noise rejection if signals V_{AC} and V_{ref} are affected similarly by the noise.

4.3.4 Synchro–Digital Converters

Among the ADCs there is a very interesting and perhaps most important family of converters made up of converters which receive information provided by a synchro or a resolver and supply digital information related to the value of the angle θ. They are divided into two categories: those which supply the values $\sin\theta$ and $\cos\theta$ in binary form, and those which calculate the value of θ and express it in binary. The converters in both these categories are designated by the same term, namely SDC, although the results of the conversions are not comparable. In practice the second category is used more often.

4.3.5   SDCs giving Sin θ and Cos θ in Digital Form

This system transforms the signals from a resolver into two pieces of digital information corresponding to sin θ and cos θ, θ being the information to be transmitted. Although they provide two pieces of information these systems are simpler than those that give the value of θ directly in digital form. In fact the angle θ is practically specified by these two pieces of information since it is easy to calculate the value of tan θ from the digital expressions of sin θ and cos θ and obtain the value of θ. Such a system uses two single phase converters which receive the information to be encoded and the same reference voltage V_{ref} as shown in the block diagram of Fig. 127. The input signals are demodulated by

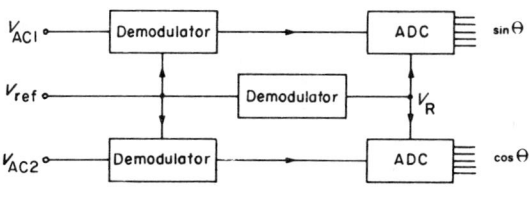

Figure 127

means of the reference voltage and fed to two ADCs which use the same external reference voltage V_R obtained by demodulating the AC reference voltage V_{ref}. Thus the ADCs give the digital information corresponding to sin θ and cos θ. As was the case for the previous arrangement this system eliminates the parasitic changes of the reference voltage. It is also possible to use only one ADC, provided both DC signals obtained by demodulating are multiplexed. The overall accuracy of the system depends upon the linearity of the demodulators, upon the quality of the multiplying coefficients which they introduce and on the accuracy of the ADCs used. In general it is not necessary to use high speed ADCs because the signal changes are usually slow compared to the conversion speed of ADCs currently manufactured. If the signals to be converted arise from a synchro-transmitter then a Scott connected transformer should be used to make them compatible with the system described.

4.3.6   SDCs giving θ in Digital Form

It is often advantageous to have θ available directly in digital form rather than use its sine and cosine to obtain it. The systems which carry out this process are sometimes called *angular position–digital* converters. Usually they incorporate a quadrant or octant selector to reduce their complexity. Before explaining the operation principle of some of the currently used systems, the operation of an octant selector will first be described.

4.3.6.1   Octant selector

This system determines the octant, or eighth part of the circle, in which the angle θ lies. To this end it determines the sign of sin θ and cos θ and compares their amplitudes. From the results of three comparisons the selector can establish the number in digital form of the octant in which the angle θ lies and gives the first three bits of the binary conversion of angle θ. These results are summarized in Table 11. From the values of V_x and V_y the selector produces two signals, V'_x and

Table 11. Results of the comparison of the signs of sin θ and cos θ to determine the octant number

Octant number	Sign of $V_x = \cos \theta$	Sign of $V_y = \sin \theta$	Comparison of the modules	Value of the most significant bits of θ
1	+	+	$V_x > V_y$	0 0 0
2	+	+	$V_x < V_y$	0 0 1
3	−	+	$V_x < V_y$	0 1 0
4	−	+	$V_x > V_y$	0 1 1
5	−	−	$V_x > V_y$	1 0 0
6	−	−	$V_x < V_y$	1 0 1
7	+	−	$V_x < V_y$	1 1 0
8	+	−	$V_x > V_y$	1 1 1

V'_y such that V'_x corresponds to the cosine of an angle β lying in the first octant and V'_y corresponds to sin β. Then the relationship between θ and β is:

$$\theta = k \frac{\pi}{4} \pm \beta$$

V'_x and V'_y satisfy the inequalities:

$$\frac{1}{\sqrt{2}} < V'_x < 1 \qquad 0 < V'_y < \frac{1}{\sqrt{2}} \qquad (V'_x > V'_y)$$

The values that must be given to V'_x and V'_y depending upon the number of the octant in which θ lies are shown in Table 12.

Figure 128 shows the schematic diagram of an octant selector. Voltages V_x and V_y are obtained by demodulating the AC voltages from the resolver. Comparators 1 to 4 receive various combinations of signals V_x and V_y and establish in which octant θ lies. The encoding logic then produces the first three bits of the encoding of θ and the control signals to the various switches which generate V'_x and V'_y from V_x, V_y, $-V_x$ and $-V_y$.

Instead of an octant selector a *quadrant selector* operating on the same principles can be used. It will simply give the first two bits of the conversion of θ

Table 12. The values of V'_x and V'_y when θ
lies in different octants

Octant	Value of V'_x	Value of V'_y
1	V_x	V_y
2	V_y	V_x
3	V_y	$-V_x$
4	$-V_x$	V_y
5	$-V_x$	$-V_y$
6	$-V_y$	$-V_x$
7	$-V_y$	V_x
8	V_x	$-V_y$

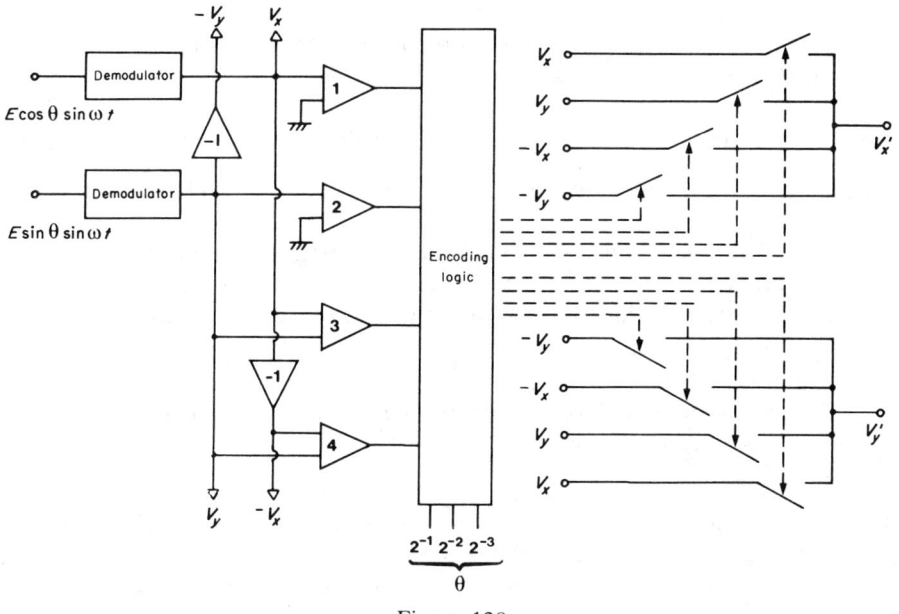

Figure 128

corresponding to the numberr of the quadrant in which θ lies. In this case the
system is simpler than the one shown in Fig. 128.

4.3.6.2 Tracking converter

In this converter the content of an n bit counter is interpreted as the binary
equivalent of an angle ϕ. After decoding, the value of this angle is compared with
the angle θ to be converted. This comparison is carried out continually. When the
angle θ varies the system produces an error signal which modifies the content of
the counter so that the angle ϕ becomes equal to the new value of the angle θ.
When a steady state is reached the contents of the counter therefore correspond to
the n bit digital word of the angle θ. Figure 129 shows the basic block diagram for

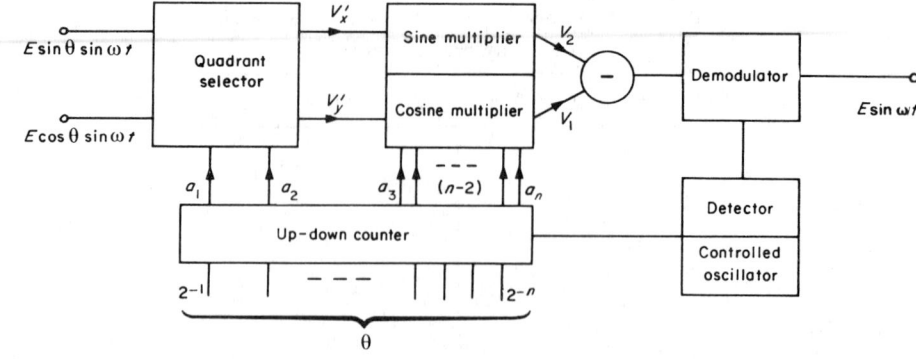

Figure 129

such a converter which will convert directly the AC signals from the resolver. Alternatively the demodulated signal may be converted, in which case there is no need of the demodulator shown in the diagram, but then two others must be placed at the inputs. The main components of the converter are:

a quadrant selector controlled by the two bits of greatest weight of an up–down counter,

two multipliers, each of which receives the remaining $(n-2)$ bits of the counter, which can be regarded as the result of the conversion of an angle smaller than 90°, and transforms this information into a sine or a cosine and multiplies it by an analog signal which it receives from the selector,

a demodulator,

a circuit for shaping the voltage error supplied by the demodulator,

a voltage controlled oscillator,

an n bit up–down counter.

The role of these components will now be examined. In the steady state the quadrant selector receives from the resolver the two signal V_x and V_y and from the counter the two bits a_1 and a_2 of greatest weight. The quadrant selector sends to the multipliers two signals V'_x and V'_y derived from V_x and V_y and proportional to the sine and cosine of an angle θ' related to θ by:

$$\theta = 180a_1 + 90a_2 + \theta'$$

Each multiplier carries out the multiplication of these signals by the sine or cosine of the angle ϕ corresponding to the $(n-2)$ bits of least weight supplied by the counter. The multiplier which receives the signal proportional to $\sin\theta'$ will carry out the multiplication by $\cos\phi'$ (and shall be called a cosine multiplier) whereas the other multiplier which multiplies a voltage proportional to $\cos\theta'$ by $\sin\phi'$ will be called a sine multiplier.

The multiplier outputs are then:

$$V_1 = K\sin\theta'\cos\phi'\sin\omega t$$

$$V_2 = K\cos\theta'\sin\phi'\sin\omega t$$

If the multipliers are not identical the constants K will be different. These two signals are subtracted from each other and the difference, which is proportional to $\sin(\theta' - \phi')$, is applied to one of the inputs of a demodulator while the reference signal $E \sin \omega t$ is applied to the other. The output is therefore a DC voltage proportional to $\sin(\theta' - \phi')$. This signal can be looked upon as an error signal since it relates the difference existing between the angle θ to be converted and the content ϕ of the counter. It is detected and used to control the frequency of a controlled oscillator. The frequency increases with the difference $(\theta' - \phi')$. The oscillator pulses are fed to the up–down counter which counts up if $(\theta' - \phi')$ is positive and counts down in $(\theta' - \phi')$ is negative. When $\theta' = \phi'$ the oscillator pulses are not sent to the counter. Therefore this system tends towards the steady state condition $\theta' = \phi'$. The result of the conversion is therefore an angle ϕ made up of the two bits controlling the quadrant selector and the next $(n-2)$ bits representing the angle ϕ'.

When angle θ changes slightly these changes appear fully in the value of the angle θ' and give rise to an error signal which in turn modifies the contents of the counter in order to eliminate this error signal.

Finally the way in which the quadrant selector is used to carry out the change from signals V_x and V_y to signals V_x' and V_y' by means of the bits a_1 and a_2 sent by the counter, must be examined. When angle ϕ is zero the first two bits a_1 and a_2 have value 0 and the selector sends signal V_x to the sine multiplier and signal V_y to the cosine multiplier and therefore $\theta = \theta'$. The demodulator output gives a signal proportional to $\sin \theta$ and depending upon its sign, the counter starts to count up or down. Consider the case when θ is located in the second quadrant. The counter counts the pulses it receives and the angle ϕ increases and the process continues as long as $(\theta - \phi)$ remains positive. When the counter has counted 2^{n-2} pulses ϕ' is again zero, the bit a_2 is 1 and then the selector changes the signals V_x' and V_y' by choosing $V_x' = V_y$ and $V_y' = -V_x$. The angle θ' associated with V_x' and V_y' is then equal to $(\theta - \pi/2)$. Now the output signal of the demodulator is proportional to $\sin(\theta' - \phi')$ and counting continuous as long as ϕ' is not equal to θ'. The accuracy of the arrangement depends upon the capacity of the counter, the accuracy of the demodulator and on the detector. Its speed depends upon the frequency of the oscillator and the response time of the analog circuits; multipliers, demodulator and quadrant selector. It is possible to improve the accuracy by replacing the detector by an integrator. In this way the oscillator frequency is increased when the difference between θ and ϕ is large, since the control voltage increases with time even if the difference $(\theta - \phi)$ diminishes. In order to avoid large fluctuations in the neighbourhood of the steady state position (which would increase the response time of the system) it is necessary that the integration should be of second order and therefore has two time constants. Since the counter can be considered to be an integrator we can take its time constant into account and therefore the integrator controlling the oscillator need only be of the first order. Such a system has the advantage of having zero speed error. In order that the system may respond rapidly to the changes of the angle θ it is necessary for the gain of the integrator and the sensitivity of the controlled oscillator to be high. This tends to promote oscillation near the steady state point, particularly quantization noise. To overcome this drawback a little hysteresis is introduced into the detector circuit, its value being adjusted to remain less than a quantum (so as not to reduce the accuracy).

4.3.6.3   Successive approximation SDCs

Some SDCs use the method of successive approximations as do a large number of ADCs. Their design follows fairly closely the design of previous converters and as was the case with ADCs the main advantage is their speed. In this type of converter the up–down counter is replaced by a sequentially addressable register. The detector is a comparator which detects the sign of $\sin(\theta - \phi)$ and when this sign is positive a pulse is sent to the register (Fig. 130). As before the quadrant selector is controlled by the two bits of greatest weight and sends to the multiplier input two signals corresponding to the sine and cosine of an angle smaller than $\pi/2$. When these two bits change their values, the two signals are modified.

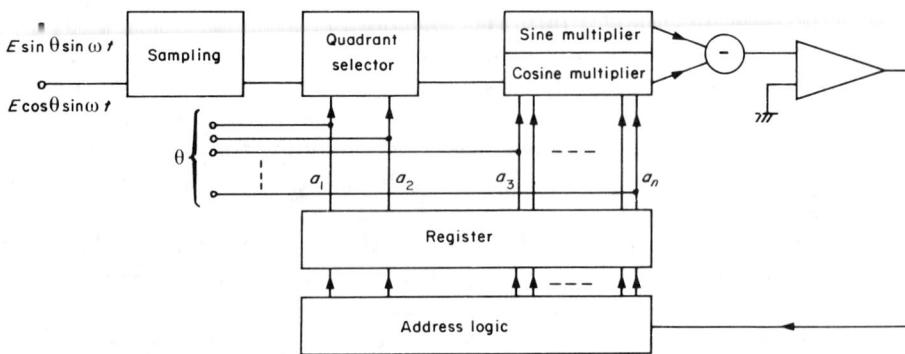

Figure 130

At the start of the conversion operation the register is reset to zero which corresponds to angle ϕ being zero. The conversion is then carried out in n identical stages of which only the first state will be described.

The bit of greatest weight of the register is made equal to 1 by the control logic. This corresponds to an angle ϕ equal to 180°. The comparator then examines the sign of $\sin(\theta - \phi)$. If θ is greater than 180° the sign is positive and the comparator sends a 1 to the control logic, which maintains the bit of greatest weight at 1. If θ is less than 180° the comparator sends a 0 to the control logic which resets the bit of greatest weight to 0 and it will then hold its value (1 or 0 depending on the result of the comparison) until the end of the conversion.

When this first stage is over the control logic makes the bit of immediately less weight a 1 and the whole procedure is started again. After each comparison the bit that has been tried remains in the state 1 or 0 depending upon the value of the comparator signal.

The various bits are thus tried one after the other until a zero signal error is obtained with the accuracy required. The value of ϕ stored in the register corresponds to the n bit conversion of the angle θ. Usually the quadrant selector is preceded by a sampling arrangement as shown in Fig. 126, and therefore there is no point in using a demodulator. A successive approximation SDC has a very high speed of conversion and good accuracy. The use of a sampling device allows it to operate with DC voltages but nevertheless it has one drawback because when the angle θ changes it introduces a speed error which varies periodically. This error which becomes zero after each sampling operation goes through a maximum given by:

$$\text{speed error (degrees)} = \frac{(d\theta/dt) \ (\text{degrees s}^{-1})}{\text{carrier frequency (Hz)}}$$

4.3.6.4 SDCs using ROMs

In contrast to the previous system, this converter generates the digital information corresponding to the angle θ using an ADC and a ROM. The main components of the system are an octant selector, an ADC which will accept an external reference and a memory which will give the value of an angle knowing its tangent.

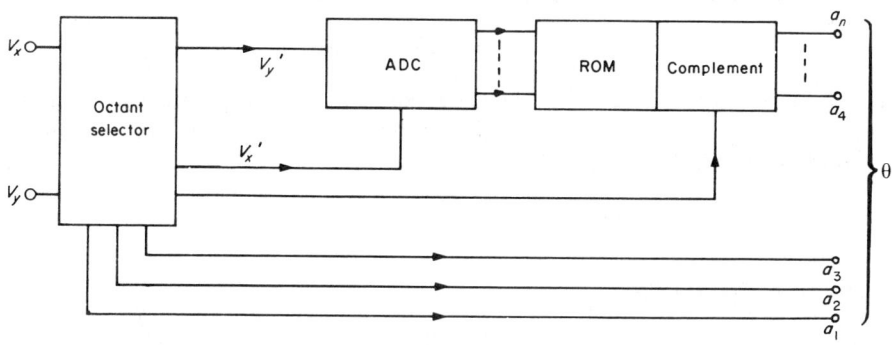

Figure 131

Figure 131 gives the basic block diagram of such a converter. The octant selector determines the number of the octant in which the angle θ lies and from this finds the first three bits of the digital word of θ. From two voltages V_x and V_y it produces two voltages V'_x and V'_y corresponding to the sine and cosine of an angle β smaller than $\pi/4$. Voltage V'_y is sent to the input of the ADC which uses V'_x as a reference voltage. The ADC thus supplies a digital message corresponding to $\tan \beta$. A ROM then gives the value of the angle β from the message it received from the ADC. The angle is obtained free of ambiguity since it is always less than $\pi/4$ $(0 \leqslant \tan \beta < 1)$. However this value of β will now always give directly the digital word for θ because the relationship between θ and β is

$$\theta = (k\pi/4) \pm \beta$$

the coefficient k being determined by the octant selector.

According to which octant the angle θ lies in a plus or minus sign will have to be taken into account. That is either the value of β is kept as it is, or its complement with respect to $\pi/4$ must be taken. This last operation is carried out in a memory location which receives from the octant selector the command to carry out or not to carry out the complement operation. In this way it provides the $(n-3)$ least significant bits of θ, the first three bits being provided by the octant selector.

A numerical example will help clarify the operation of this converter. Let it be assumed that an angle $\theta = 160°$ is to be encoded using seven bits and that the two voltages V_x and V_y obtained by sampling are:

$$V_x = -0.87K \quad \text{and} \quad V_y = 0.33K$$

K being a scale factor.

The octant selector finds out that θ is located in the fourth octant and hence generates the first three bits having the value 011. The octant selector also sent the signal V'_x and V'_y such that:

$$V'_x = 0.87K \quad \text{and} \quad V'_y = 0.33K$$

which correspond to the cosine and sine of angle $\beta = 20°$. The ROM supplies the four bits corresponding to this angle, that is 0111 (the quantum has the value $2°.81$). The angle θ to be encoded can be expressed as:

$$\theta = 160° = 3 \times 45° + 25° = 4 \times 45° - 20°.$$

In order to express θ in terms of the coded angle β, the complement of β relative to 45° must first be obtained which gives 1000. The binary word for the value of the angle θ is therefore 0111000.

4.4 Examples

Tables 13 and 14 give examples of the characteristics of some synchro–digital and digital–synchro converters.

Table 13. Characteristics of two synchro–digital converters

	Manufacturer (type)	
	Analog Devices (SDC 1704)	DDC (HSDC14)
Resolution (bits)	14	14
Accuracy	$\pm 2.2'$ (400 Hz signal)	$\pm 4'$
Input signals and corresponding impedances	11.8 V_{rms} 26 kΩ 90 V_{rms} 200 kΩ	11.8 V_{rms} 17 kΩ 90 V_{rms} 130 kΩ
Reference voltage	26 V_{rms} or 115 V_{rms}	26 V_{rms} or 115 V_{rms}
Output signals	TTL compatible natural binary code	TTL/CMOS compatible natural binary code
Response to a step of 179° (ms)	125	120
Maximum possible speed variation of input signal (revolution s^{-1})	12	10^{-1}
Maximum possible acceleration (°s^{-2})	36 000	58 000
Supply	+15 V 30 mA −15 V 30 mA 5 V 85 mA	+15 V 40 mA −15 V 15 mA 4.5–15 V $R_e = 10$ kΩ
Special facilities	A voltage proportional to the angular speed of the input signal is available	Voltages proportional to the angular speed of the input signal and also to difference input/output are available

Table 14. Characteristics of two digital–synchro converters

	Type	
	DSC 801 (CCC)	EDSC (DDC)
Resolution (bits)	14	14
Input data	Natural binary code Parallel data Level DTL/TTL	Natural binary code Parallel data Level DTL/TTL
Output signal for synchros	11.8 V_{rms} in 150 Ω 90 V_{rms} in 5 kΩ	11.8 V_{rms} in 100 Ω 90 V_{rms} in 5 kΩ
Reference voltage (V_{rms})	26 115	26 115
Conversion speed (μs)	50	—

5 COMPONENTS USED IN CONVERTERS

Over the last twenty years a real revolution has taken place in the method of manufacturing converters. Twenty years ago a ten bit ADC made of discrete components dissipated 500 W and required a volume of several hundred dm^3. Nowadays converters having the same resolution dissipate less than 100 mW and are available in 16 or 24 pin integrated circuit packages. These enormous savings of space and power have been achieved thanks to integrated circuit techniques. Thus a ten bit ADC will group on one chip of 2×2.3 mm, a reference source, ten transistors used as current sources, a reference amplifier, an offset current source for bipolar operation and a thin film resistance network having an equivalent value of 1.5 MΩ. However these new fabrication techniques give rise to new problems. Thus the components are much more closely packed than when discrete components were used and the temperature variations that are likely to occur will have greater effects upon the system performance. On the other hand the accuracy demanded means that for a twelve bit converter the fabrication technology is stretched to the very boundaries of present knowledge. Therefore the fabrication of converters using these new techniques is a very complex operation and the very high resolution circuits (corresponding to 14 or 16 bits) are still often manufactured as hybrid circuits. It is therefore of interest to examine the influence of integrated circuit technology upon the four main components mentioned above.[54]

5.1 COMPARATORS

5.1.1 Definition of a Comparator

One of the most important operations carried out during the quantization of analog signals is the *comparison* of two voltages. Any analog to digital converter has at least one comparator whose performance determines to a large extent the performance of the ADC using it. The comparator[55] can be said to be a hybrid module at the boundary where analog and digital techniques meet. A comparator is a device with two inputs and one output, similar to an operational amplifier, moreover they are often represented by the same symbol. Its function is to provide a digital output signal N which can only have *two values* (corresponding to two states 0 and 1) depending upon the relation existing between the magnitudes of the two input voltages V_1 and V_2. For instance, if V_1 is greater than V_2, N corresponds to the state 1 and if V_1 were less than V_2, N would correspond to the state 0. In order to fulfil this function any comparator must have a certain number of circuits as shown in Fig. 132. Essentially, these are *subtracting circuits*, which

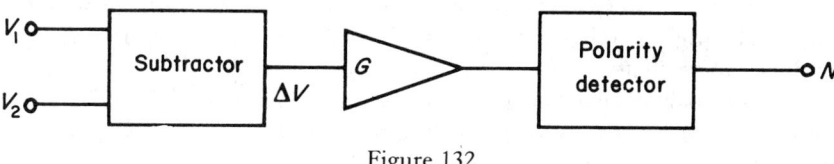

Figure 132

obtain the difference $V_1 - V_2 = \Delta V$, *an amplifier*, which amplifies that difference, a circuit which *detects the polarity* of the amplified signal and an *output stage* which shapes the digital output signal N. The gain G of the amplifier must be adequate so as to ensure that the voltage $G \Delta V$ is greater than the detector threshold. Figure 133 shows the ideal transfer characteristic of a comparator. In practice

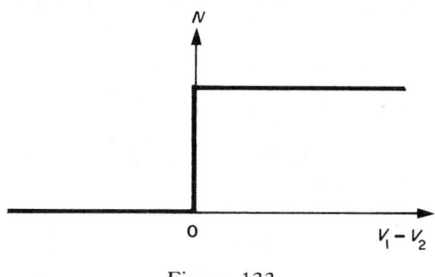

Figure 133

dead zones are present, that is there is a set of values of $(V_1 - V_2)$ for which the state of the output signal is not defined. This dead zone is centred around $V_1 - V_2 = 0$. It can then be said that within this dead zone the comparator operates as a linear amplifier and that outside that zone the amplifier is cut off or saturated. It is, moreover, usual to consider the comparator as being an operational amplifier with specific characteristics.

5.1.2 The Difference Between an Operational Amplifier and a Comparator

It is sometimes possible to use an operational amplifier as a comparator, in particular for some low frequency applications. There are, however, differences between these two types of device which need to be clarified.

The output stages of comparators are designed so that their circuits are compatible with the usual logic circuits used (TTL in particular). They can control several circuits and this requires a large output current.

Comparators have wider pass-bands and shorter propagation times than amplifiers.

Comparators usually have a greater input dynamic range than amplifiers whether for the common or differential mode configuration.

In contrast to operational amplifiers, comparators are designed to operate with an open feedback loop (with no negative feedback). Since they have a wide pass band and a high gain in order to be sufficiently fast, they have an unfortunate tendency to oscillate.

Comparators, particularly the fastest ones, have lower input impedances and higher bias currents than operational amplifiers and this is normally accompanied by significant offset voltage and current drifts.

5.1.3 Types of Comparators

There are mainly two ways of using a comparator, i.e. depending on whether the signals to be compared are *currents* or *voltages*. For the comparison of voltages the two usable signals are applied directly to the inverting and non-inverting inputs of the comparator (Fig. 134), and the output signal N indicates immediately the sign

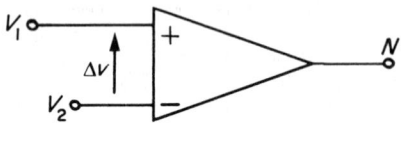

Figure 134

of the difference $V_1 - V_2 = \Delta V$. In the case of an ADC signal, V_1 corresponds to the signal to be converted whereas the voltage V_2 is the signal produced by a DAC or a ramp generator. In order to be used in this configuration the comparators must have an excellent common mode rejection and accept a large variation of the common mode signal. For example, a twelve bit ADC must have a common mode rejection factor of 96 dB if the common mode error is required to be less than a tenth of a quantum. Usually, in order to minimize these errors, steps are taken to reduce the dynamic range of the input signals to the comparator which requires that the dead zone be reduced and consequently the gain increased.

It is also possible to carry out the comparison of currents using the circuit of Fig. 135. This is particularly useful if the reference source generates a current (the case of a DAC with current output). The output signal changes state when the voltage V_1 passes through the value

$$V_1 = R_1 I_{\text{ref}}$$

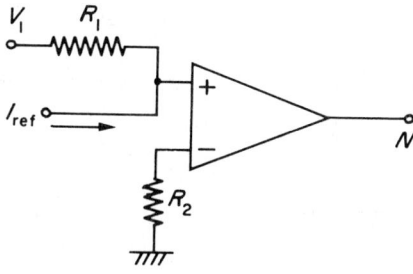

Figure 135

The resistor R_2 is for compensating the offset currents. Moreover, with this circuit greater switching speeds are obtained.

5.1.4 Comparator Characteristics

An ideal comparator would have the following characteristics: zero current and voltage offset, infinite gain, infinite input impedance and infinite pass band, that is, zero comparison time. In practice the performance of comparators is limited by parameters similar to those encountered with operational amplifiers, namely: open loop gain, pass band and finite rate of changes of the output signal. The effects of these parameters upon the transfer characteristic, the gain and on the speed must therefore be examined.

   The comparator specifications supplied by a manufacturer are not always easy to understand because the meaning of the terms used is not necessarily the same as for operational amplifiers. It also happens that widely different terms are used to describe identical phenomena. For example one speaks of the propagation time of a comparator and of the pass band or of the rate of change of the amplifier signal. Finally the methods of measurement used by manufacturers are not always the same and this may give different results for the same comparator. It is therefore important to read carefully the manufacturer's specifications.

5.1.4.1 Ideal transfer characteristic

In practice the transfer characteristic of a comparator resembles the curve shown on Fig. 136. This curve exhibits a region called *dead zone* or *window* in which the output signal takes neither of the two values corresponding to the states 0 or 1. The width of this zone, in which the comparator operates linearly, depends on its open loop gain. For example, if the dynamic range of the applied signal is 10 V and an accuracy of 12 bits is required for the conversion, it is necessary that the width of the window does not exceed 1 mV (which corresponds to a half quantum). If the difference between the two values of N is 3 V (TTL compatibility) this requires that the comparator open loop gain be at least 3000. Usually the gain is higher than this and the window reduced accordingly.

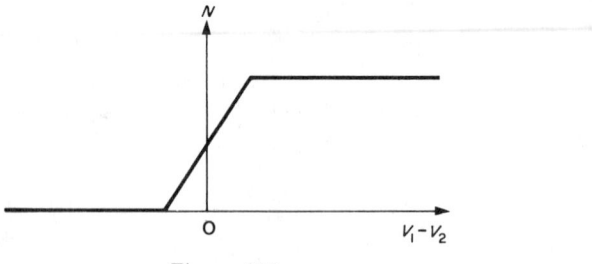

Figure 136

5.1.4.2 Accuracy

The accuracy of a comparator depends mainly on two factors; the *real input offset voltage* and the error due to the *limitation of the gain* (this last factor results from the existence of the dead zone or window). With the very high gain comparators currently available the second cause of error can be neglected (for an output voltage of 5 V, a gain of 100 000 results in an error of 50 μV). In this case the offset voltage mentioned corresponds to the definition used for operational amplifiers. When the gain is low this error is often separated. This occurs particularly for very fast comparators, since a good accuracy (that is, a high gain) and high speed are two criteria difficult to reconcile. Usually the two causes of error are grouped together by the manufacturers under the name of offset voltage which introduces a risk of confusion. As with operational amplifiers it is possible to eliminate the offset voltage by means of an opposite voltage which is adjusted by means of an external potentiometer. However, the offset voltage changes with temperature and this drift cannot be compensated for. Therefore the change of offset voltage with temperature is a very important parameter even though it is not always specified by the manufacturers. In ADCs it is included in the changes of the offset error (or of the zero) with temperature. The accuracy also depends upon the common mode voltage and on the load impedance which affects the value of offset voltage.

5.1.4.3 Speed

Another important parameter of a comparator is its speed which is usually defined as the time between the instant of excitation and the time when the output signal reaches 50% of its final value (Fig. 137). This time is called the *response time* or

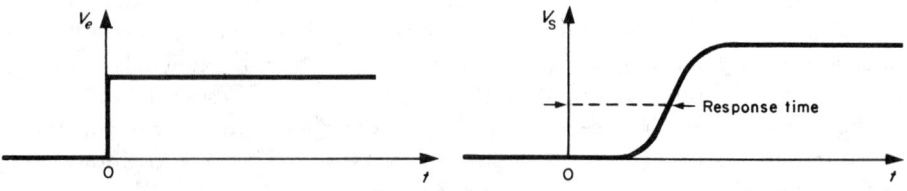

Figure 137

propagation time. The value of the load resistance and the values of the logic levels affect this response time. The amplitude of the value applied between the two inputs also plays an important role.

The response time will vary greatly depending on whether the comparator remains in the zone of linear operation or is saturated by the input signal. In the latter case the comparator is *overdriven* and the greater the overdrive the shorter the response time will be.

The method usually used is as follows. A differential voltage of 100 mV (positive or negative) is applied between the two inputs. The comparator will then be in a well defined state (0 or 1 depending upon the sign of the applied voltage). A voltage of opposite sign is then applied to it having a value of 100 mV plus the overdrive. The response times are slightly different according to whether positive or negative input signals are used. Figure 138 reproduces the response curves for

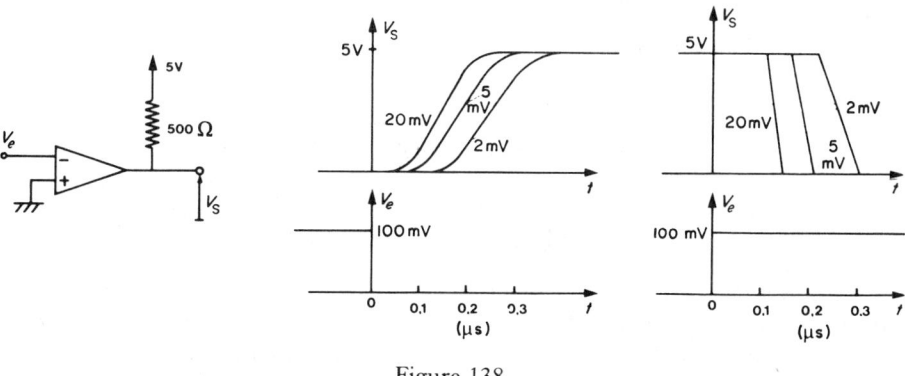

Figure 138

various overdrives for the LM311 comparator, manufactured by National Semi-conductor. In order to be able to use these results, it is very important to take into consideration the load resistance used by the manufacturer and we have therefore also included the circuit used to carry out the measurements.

5.1.4.4 Stability

A comparator is a high gain system with no feed back and therefore it has a tendency to *oscillate* when it is operating in its linear part. This behaviour can have serious consequences for the operation of an ADC, for if the comparator tends to oscillate when the signal is nearing the limit of a quantization level, the output code will no longer be defined and the ADC behaviour may be non-monotonic. Hence certain precautions must be taken to avoid these oscillations and to this end the comparator may be isolated from its load by using a transistor in the common-collector configuration. It is important to note that the higher the comparator speed the greater are the risks of oscillation occurring.

High input impedance, possible large variations of the load, high speed of operation and high accuracy are all parameters which are not easily compatible and it is therefore not always advisable to look for a comparator having all these characteristics as it will entail a high price and difficulties in its design and manufacture. Consequently comparators have been divided into two categories; *fast comparators*, and *accurate comparators*.

As an example, Table 15 gives the main characteristics of the LM311 comparator made by National Semiconductor.

**Table 15. Some characteristics of the National Semi-
conductor LM311 comparator ($T_a = 25\,°C$)**

	Typical value	Maximum value
Offset voltage at input, $R_S \leqslant 50\,k\Omega$ (mV)	2	7.5
Offset current at input (nA)	6	50
Bias current (nA)	100	250
Voltage gain	200 000	
Response time[a] (ns)	200	

a) Overdrive of 5 mV. Connection of Fig. 138.

5.1.4.5 Technology of production

The techniques used for the production of comparators are very much those used for the manufacture of operational amplifiers. Nowadays the tendency is towards the use of monolithic integrated circuits, the only limitations arising because of the switching speeds. We shall limit ourselves to a brief description of a comparator used in a high performance ADC (6 bits, 5 ns).

It is made up of an input operational amplifier followed by a flip-flop for reshaping the output and an attendant current switching system to pass from the amplifier mode to the memory mode (Fig. 139). The input amplifier consists of transistors T_1 and T_2 in the common collector configuration and of transistors T_3 and T_4 of the differential stage. The flip-flop consists of transistors T_7 and T_8. Transistors T_9 and T_{10} mounted in the common-collector configuration provide the power gain during the reshaping period and isolate the output. Transistors T_5, T_6, T_{12} and T_{18} act as current switches between the amplifier and the flip-flop. When the potential of input U exceeds the potential of input L the signals from the differential stage are applied to resistors R_1 and R_2 and a potential difference appears between the bases of transistors T_7 and T_8 of the flip-flop which is then not energized. Then, when the potential of input L exceeds the potential of input U the currents from the differential stage go through transistors T_{12} and T_{18} which put the flip-flop into the required state (0 or 1).

The rest of the comparator consists of current sources for the amplifier and reference voltages for the various stages. The characteristics obtained for this comparator are as follows; offset voltage = 2 mV; common-mode rejection >70 dB; hysteresis <1.5 mV; response time 3.5 ns.

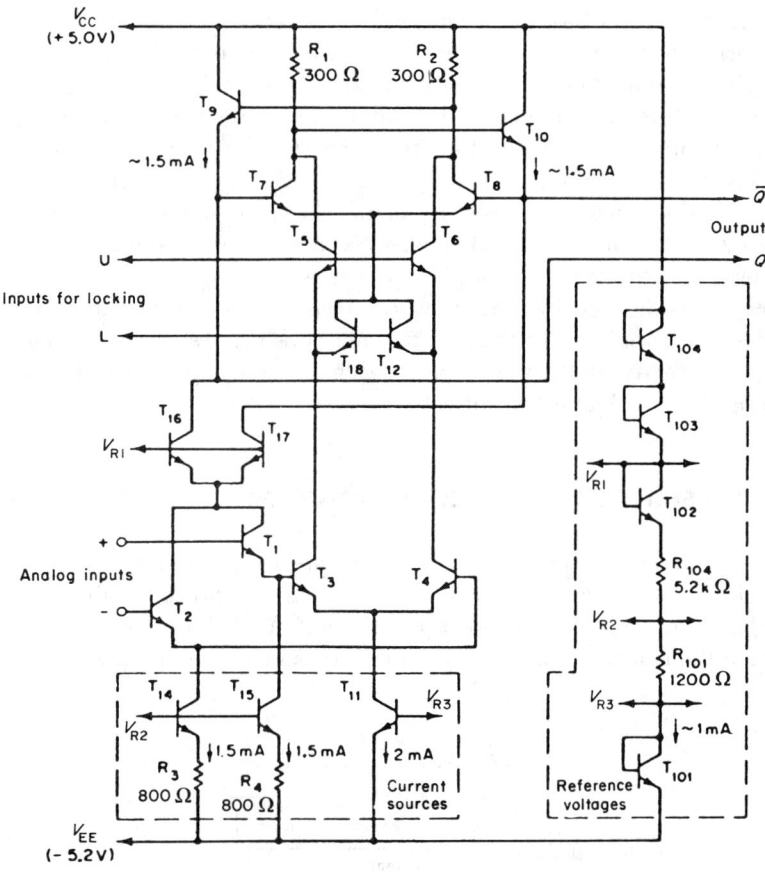

Figure 139

5.2 RESISTANCE NETWORKS

In 1970 most manufacturers and users were predicting that it would be impossible to fabricate monolithic ADCs with a resolution greater than 7 or 8 bits. According to them this limit was imposed by the stability that could be achieved with integrated resistance networks.[54] Nowadays several manufacturers are offering 10 or even 12 bit integrated circuit (i.c.) ADCs.

In order to better appreciate the problems that had to be overcome and to obtain an idea of the ultimate performance that can be expected, one must be reminded of the accuracy demanded of the components of a ten bit ADC. The overall error must be less than a half-quantum, that is 0.5×10^{-3}. This determines the accuracy that the network resistors used must have. The ratios between these resistors is of the order of 10^{-4}. Moreover if it is required to operate within the temperature range of 0–70 °C then the overall variation must be at most 10^{-4}, which corresponds to a temperature coefficient of about $10^{-6}\,°C^{-1}$. These few figures give an indication of the problems to be faced. Before starting to compare the various fabrication processes it is important to be reminded that these i.c. methods do not produce resistors that are systematically better than those obtained with discrete components. Their main advantage lies in the considerable space saving that can be achieved. In general, the values of resistors used in ADCs are high (10 kΩ and over) and to achieve a substantial saving of space the fabrication process must allow high resistivities.

Three methods used for obtaining i.c. resistors will be explained and their performances compared. In the first two methods the resistors are produced by the same methods used for the fabrication of semiconductors and the third method is thin- or thick-film fabrication.

5.2.1 Resistors Obtained by Diffusion

Resistors obtained by diffusion belong to the category of semiconductor resistors. This process is the least complicated since it requires no extra fabrication stage other than those used for bipolar transistors. A diffused, resistive structure consists of the resistance offered by a diffused semiconductor region. Figure 140

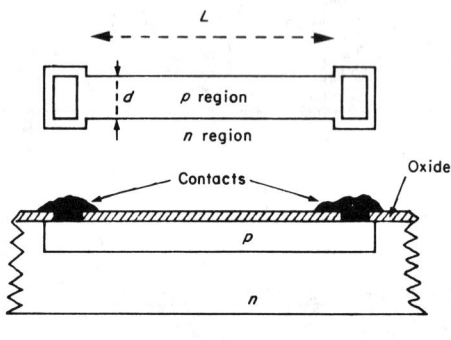

Figure 140

shows the structure geometry and the cross-section of a p type diffused resistor. When such a structure is used it is necessary to reverse-bias the p-n junction thus obtained in order that all the current flows through the resistive part. The value of the resistance may be given by:

$$R = \rho(L/S) = \rho(L/dx)$$

in which x is equal to the depth of the junction. In practice the parameter R_s is used:

$$R_s = \rho/x$$

R_s is the *surface resistance* and has the dimension of resistance and corresponds to the resistance per unit area $\Omega/\square$ (the $\square$ symbol is a reminder that it is a resistance per unit area). Typical values obtained by diffusion range from 60 to 250 $\Omega/\square$ and the resistors have a high temperature coefficient (of the order of 1000 to 2000 ppm °C^{-1}). These resistors are seldom used in DACs because of the high temperature coefficient and of the fact that it is difficult to exceed 200 $\Omega/\square$.

5.2.2 Use of Ionic Implantation

The second method of obtaining i.c. resistors is by ionic implantation. In this technique, the impurities creating the resistivity are introduced into the crystalline structure of silicon by bombarding the surface of the material with high energy ions. The implanted ions form a very thin layer at the silicon surface (typically of the order of 0.5 nm). Thus for equivalent doping levels ionic implantation gives a surface resistance 20 times greater than the resistance of a diffused layer. The surface resistance of resistors obtained by ionic implantation is inversely proportional to the dose implanted. By proper choice of the dose the surface resistance can be varied from 200 $\Omega/\square$ to 20 k$\Omega/\square$ and the values can be controlled very precisely. Thus it is possible to obtain very well defined resistance values and very good matching of resistors.

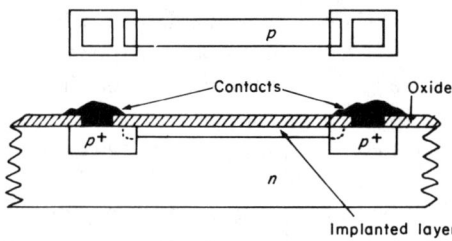

Figure 141

Figure 141 shows the structure lay-out and the cross section of a p type resistor obtained by ionic implantation. As the resistor is very thin it is difficult to make good ohmic contacts on the implanted region. It is therefore necessary to prepare two diffused p regions which will serve as contacts and since the resistance value of these regions is much smaller than that of the implanted region the actual value of the resistance will be defined by the implantation process only. The temperature coefficient of this type of resistor is about one quarter of the coefficient obtained for diffused resistors. In contrast to thin film resistors, diffused and ionic implantation resistors are protected by a thin oxide layer. Finally this system does not require high fabrication temperatures and therefore the semiconductor parts can first be fabricated by standard diffusion processes and then the required resistors can be added, without causing any alteration to the diffused parts.

5.2.3   Film Resistors

The third possible method, tending nowadays to become a general method, consists in depositing the resistive material as a *thick* or *thin film*, there being no decisive advantage for one or the other type. Thin film resistors are used when great accuracy and good stability are required; on the other hand thick film resistors have better power ratings and cost less.[57-59]

Thin film resistors have a thickness of less than a micron, typically from 10 to 100 nm. They are produced by evaporation under vacuum or by cathodic bombardment. The metal which forms the resistance is deposited directly on to the silicon substrate or on an intervening oxide layer (Fig. 142). Usually nichrome

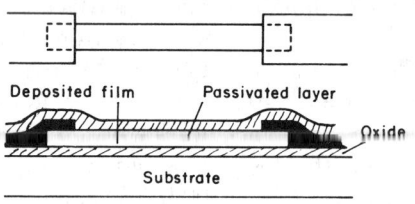

Figure 142

(nickel–chrome) or a derivative of tantalum (in particular tantalum nitrate Ta_2N) is used. The deposition of the metal can be done in several ways. Firstly metallic or plastic masks can be used, the metal being deposited through the slits in the masks. It is also possible to deposit the metal over all the substrate surface and only keep the useful parts by using a photosensitive resin which is illuminated through a mask. Thin film resistors are surface devices and it is necessary to carry out a thorough passivation in order to ensure long term stability. When they are used in a DAC it is advisable to carry out two passivations. This thin film technique gives easily reproducible resistance values because the film thickness which is the main parameter in this method of fabrication, can be controlled very accurately for each batch. The accuracy for thin film resistors is usually of the order of 5%.

In contrast to thin films, *thick film resistors* have a thickness exceeding 10 μm. A past or ink is used to make them which is applied to the substrate through a silk screen. This silk screen has areas or windows through which the paste flows and then the assembly is baked in an oven at between 800 and 1000 °C to ensure that film adheres well to the substrate. The material most used for substrate is alumina. The resistive ink is made of a metal or a metallic oxide mixed with glass and a resin is added as binding agent. By varying the proportion of this mix it is possible to obtain a large range of resistance values. Baking modifies the resistivity of the mix and hence the value of the resistances can vary from their initial value by 50% so that an adjustment is necessary.

The advantages of both techniques are compared in Table 16, while Table 17 gives the performance that can be obtained.

Table 16. Advantages of thin and thick film techniques for making resistors

	Thin film	Thick film
Size of circuits	small	small
Tolerance	excellent	good
Temperature coefficient	excellent	good
Resistivity range	good	very wide
Resistance range	good	excellent

Table 17. The performance of thick and thin film resistors

	Thin film	Thick film
Resistivity range (kΩ/$\square$)	0.05–0.5	15–330
Resistance range (Ω)	10–10^7	5–10^8
Temperature coefficient (ppm °C^{-1})	±20–±100	±200–±300
Equality of 2 coefficient (ppm °C^{-1})	±1	±5–±50
Resistance tolerance (%)	±0.001–10	±0.5–±20
Matching of 2 resistances (%)	±0.01	±0.05–±0.5

With both techniques the value of the resistances is determined by the resistivity of the film; its thickness is not taken as a variable. Although the tolerance and stability of film resistors are comparable to those of discrete resistors the former are superior, above all because of the precise ratios obtainable and the equality of the two temperature coefficients. However compatibility with monolithic circuits depends on their structure and dimensions so preference is given to thin films while thick films are used to a greater extent in hybrid structures.

5.2.4 Comparison of the Methods

The three methods give integrated circuit resistors which differ in the complexity of the processes used to make them. The simplest system is the diffusion process which does not require any extra stage other than those used for the fabrication of semiconductor components.

Ionic implantation is of great interest since it does not need a high temperature stage so that earlier diffusion stages are not affected. The fabrication of thin film resistors raises the problem of ohmic contacts, but they are not subject to the same lay-out constraints as the resistors obtained by the two previous methods. The ionic implantation and the thin film methods offer better temperature stability and a resistivity ten times higher than that obtained by the diffusion method, though this last method is of interest because of its lower price. Table 18 reproduces the comparative data obtained for the three methods by Signetics.[26]

In fact whatever the method used to fabricate a resistance network the temperature coefficients are much more likely to be similar than if discrete

Table 18. Comparison of three ways of making integrated circuit resistors

Fabrication process	Diffusion	Thin film	Ionic implantation
Nominal resistivity ($\Omega/\square$)	135	1000	1250
Tolerance on matching (%)	0.07	0.06	0.05
Temperature coefficient (ppm °C^{-1})	1500	−200	400

components were to be used. This is mainly due to the fact that the resistors of the network are made of the same material, that they are very close to each other on the substrate and that the substrate ensures a good temperature distribution. The three processes give rise to identical problems. First the resistors obtained at the end of the fabrication process do not have the required resistance values (the variation can be of 50%). It is then necessary to adjust them and this is usually done by modifying the area of the resistors by means of a laser beam. In order to obtain a uniform layer it is necessary that the substrate be as flat as possible and this is particularly important for thin films for which the thickness of the resistive part is of the order of 0.1 μm. Finally the state of the edges also has a great influence, since the resistance is proportional to the area of the material deposited. Hence masks of high quality with well finished edges must be used. Table 19 gives the performance achieved by Analog Devices and Hybrid Systems with thin film techniques.

Table 19. Performance of resistors made with thin film techniques

	Analog devices		Hybrid systems	
	Typical value	Maximum value	Typical value	Maximum value
Resistivity ($\Omega/\square$)	5–500		300	
Resistance tolerance (%)	0.001	20	0.1	1
Tolerance on ratios	—	—	0.01	1
Temperature coefficient (ppm °C^{-1})	±20	±100	±35	±50
Matching of 2 coefficients (ppm °C^{-1})		±1		±2
Absolute stability (ppm yr^{-1})	75	100	±500	±1000
Stability of ratios (ppm yr^{-1})	20	50	±50	±100

5.3 ANALOG SWITCHES

5.3.1 General

An analog switch[4] is a device which will allow an analog signal through (ON state) or block it off (OFF state). An ideal analog switch must therefore behave as a

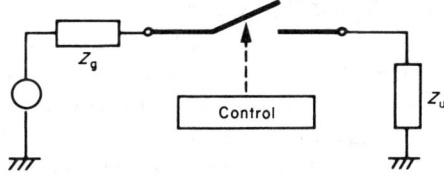

Figure 143

short-circuit or an open-circuit depending upon its state, the change of states being controlled by a control circuit dissipating no power (Fig. 143).

The currently used switches use *semiconductors* (bipolar transistors or FETs). When it is conducting the switch must have a low resistance, a small offset voltage (or saturation) and the capacity for switching bipolar signals of large amplitudes. In the OFF state it must be similar to an infinite impedance. It must have a high switching speed requiring very little power for its control.

FETs are more useful than bipolar transistors for this application because they do not display any residual voltage, they isolate the control signal very effectively from the signal to be switched and lend themselves to integrated circuit fabri- cation. However, their switching speed is less than that of the bipolar transistor. Since the signal to be switched can be either a voltage or a current the switches can be divided into two categories: *current switches* and *voltage switches*. In DACs one or the other category is used depending upon the circuit. For example a ladder DAC will use voltage switches but an inverted ladder or weighted resistor DAC will use current switches. Current switches are generally easier to produce because transistors lend themselves better to current switching than to voltage switching. Nevertheless problems arise in either case as soon as resolutions of twelve bits are required.

There is another method of classifying the switches. One can distinguish *switches* (as shown in Fig. 143) from *change-over switches*. In the case of an inverter, two signals V_1 and V_2 from two generators V_1 and V_2 may be connected to the same output (Fig. 144). The term commutator is sometimes used. These

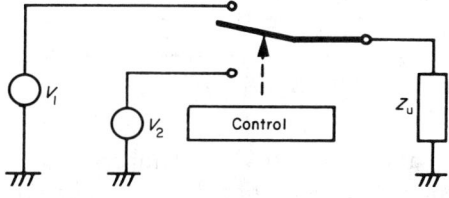

Figure 144

two categories are usually designated by the following abbreviations; SPST (single pole single throw) for a switch, and SPDT (single pole double throw) for a change-over switch. From now on, when necessary, a switch will be clearly specified as a switch or a change-over switch.

5.3.2 Bipolar Transistor Switches

The first analog switches made used bipolar transistors. In order to use a bipolar transistor as an analog switch its base-emitter junction must alternately be OFF and ON. In this last state it is possible to have a very low collector to base voltage (designated V_{CEsat}) if the base current is of the order of one tenth of the emitter current and if the emitter current itself is sufficiently low. This voltage V_{CEsat} corresponds to the offset voltage of the switch. Offset voltages of the order 10 mV can be obtained for saturation currents of 1 mA. In this case the equivalent dynamic resistance of the transistor is of the order of 30 Ω.

These figures are too high if it is required to carry out a 10 or 12 bit conversion. In order to reduce them, the inverted connection is used (Fig. 145) in which the

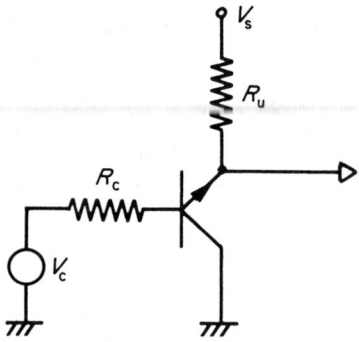

Figure 145

emitter and collector are interchanged. If the control signal V_C is positive the transistor is ON, whereas when V_C is more negative than V_s the transistor is OFF. In this configuration the current gain is greatly reduced (it can even become less than unity), and the saturation voltage can become less than a millivolt and is given by:

$$V_{EC} = \frac{kT}{q} \ln \frac{\alpha_N \left(1 - \frac{1-\alpha_I}{\alpha_I} \cdot \frac{I_E}{I_B}\right)}{1 + (1-\alpha_N)\frac{I_E}{I_B}} + r_C I_B$$

where k is the Boltzmann constant, T the temperature in K, α_I and α_N are the direct and inverse current amplification factors, and r_C the ohmic resistance of the collector (proportional to $1/I_B$). When the ratio I_E/I_B is sufficiently small V_{EC} can be written:

$$V_{EC} = \frac{kT}{q} \ln \alpha_N + r_C I_B$$

The saturation resistance can be of the order of a few ohms only.

The second simple circuit using bipolar transistors is the series–parallel or change-over switch. It consists of two mutually exclusive switches, each switch using an *npn* or *pnp* transistor (Fig. 146). It is not possible to use the same source for controlling the two transistors because they require currents of opposite sign. The layouts used differ in the way the base driving circuits are made. Figure 147 shows the connection diagram for such an arrangement, which requires an auxiliary voltage source which is greater than the voltage to be switched V_{ref}. These arrangements use an appreciable number of components and their translation into monolithic fabrication raises several problems. There are not many integrated circuit switches of this type available commercially.

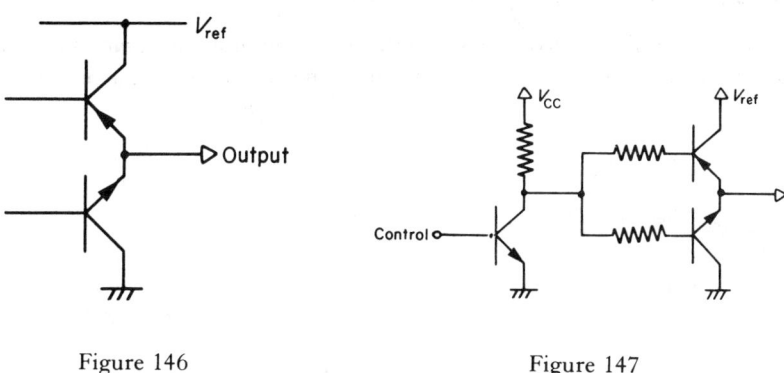

Figure 146   Figure 147

Current switches are usually based on the principles of Fig. 145, whereas change-over switches are often used as voltage switches.

5.3.3 FETs as Switches

Junction or MOS transistors are being used increasingly in switches because of their low power dissipation and the ease of integration. Whatever type of transistor is used the two parameters of importance are the equivalent transistor resistance in its conducting state and the amplitude of the analog signal that can be switched.

5.3.3.1 Junction gate FET

This is a *depletion device* in which the channel always exists between the source and the drain. In the absence of bias the transistor is conducting (ON). The gate–source junction is usually reverse biased and the channel conductance varies linearly with the voltage V_{GS} (Fig. 148 shows one n channel).

When the gate–source junction is forward biased the conductance increases but this working condition corresponds to a significant control current which may distort the signal to be transmitted.

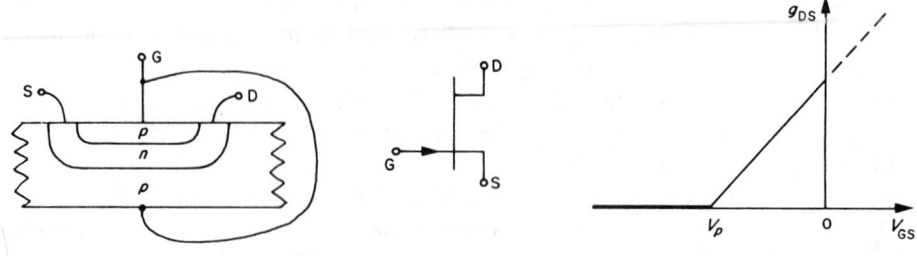

Figure 148

5.3.3.2 Insulated gate FET (MOS)

The most useful insulated gate transistors are *enhancement devices* and the channel
does not exist at rest (Fig. 149) only appearing when the gate is biased. Hence the

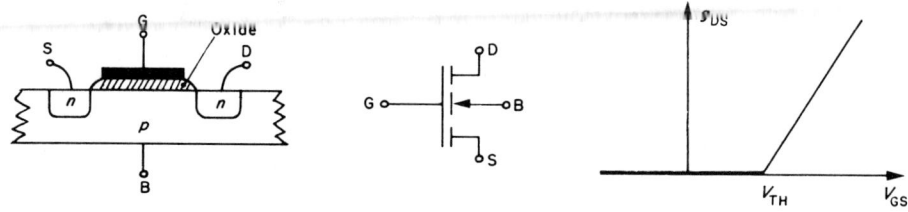

Figure 149

device is turned OFF in the absence of biasing and the voltage corresponding to
the limiting cut off is the threshold voltage V_{TH}. Insulated gate transistors are in
practice four terminal devices, the fourth terminal being the substrate B. The
substrate and the *n*-regions form *p-n* junctions which must always be reverse
biased and hence terminal B is connected to the most positive potential in the case
of a *p* channel and to the most negative potential in the case of an *n* channel.

5.3.3.3 Comparison

For analog switching it is advantageous to have the highest conductance possible.
The dimensions of the channel can be increased but this is limited because an
increase of the channel width results in an increase in the parasitic capacitances
and in the leakage currents. For equal geometries, junction FETs have a lower
conductance than MOS devices. Moreover *n* channel devices have a higher
conductance than *p* channel devices as the electron mobility is greater than that of
the holes.

   Two phenomena limit the operation of an FET used for switching, the
switching time and the *transient peaks* appearing in the logic circuit which are due
to the control signal entering the analog circuit. The switching time can be
estimated using the equivalent circuit representation of the FET, but these
calculations are difficult to carry out because the capacitances involved depend

upon the values of the control signal and the signal to be switched. The switching time can be reduced by using control signals of smaller amplitudes but this also reduces the amplitude of the signals that can be switched.

5.3.4 Junction FET Switches

Most junction FET switches are n channel devices. When an analog signal is to be switched with this type of transistor two problems arise; the signal to be switched is usually a bipolar signal of fairly large amplitude (10 V) whereas the control logic signal has an amplitude of the order of 5 V (TTL compatibility), and the gate–source junction must always be reverse biased.

In order to resolve these two problems the control circuit is placed between the logic input and the transistor gate so that in the conducting state the gate potential follows the source potential in order to ensure that:

$$V_P < V_{GS} < 0$$

There are a variety of control circuits because there are several ways of ensuring that the gate potential follows the source potential. Diodes and RC circuits can be used, or junction transistors or even FETs. Technically these are considered to be

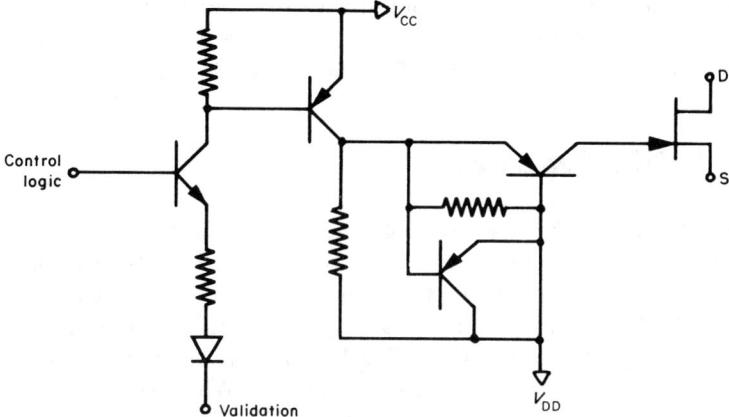

Figure 150

hybrid circuits since the FET chip is connected to the integrated circuit. Figure 150 shows, as an example, the circuit diagram of the switch DG 140 A manufactured by Siliconix. Its characteristics are as follows:

drain-source resistance in the conducting state $r_{DS}(ON) = 10\ \Omega$,
OFF–ON switching time $t_{ON} = 0.5\ \mu s$
ON–OFF switching time $t_{OFF} = 1.25\ \mu s$
ON state control power $P_{ON} = 175\ mW$.

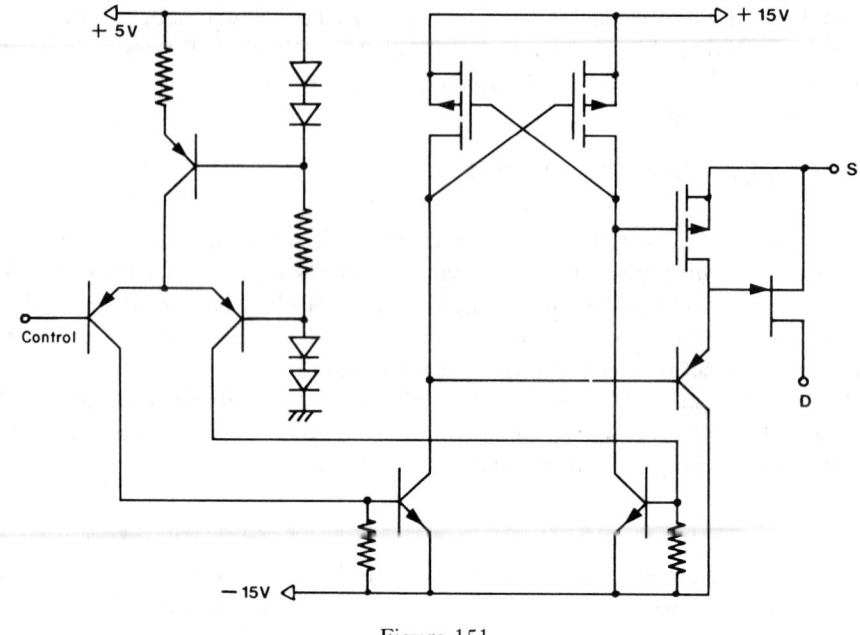

Figure 151

The circuit diagram of Fig. 151 is for the Siliconix switch DG 181, and the control arrangement uses MOS devices in order to reduce the power dissipated. This circuit gives an r_{ON} of the order of 50 Ω and a switching time t_{ON} of 150 ns.

5.3.5   CMOS Switches

Enhancement MOSs have a *threshold* that has to be exceeded in order that the device becomes conducting. This threshold limits the amplitude of the signal that can be switched and moreover the resistance presented by the device varies greatly with the amplitude of the analog signal to be switched. These two problems are solved by using CMOS technology in which an *n* channel MOS is connected in parallel with a *p* channel MOS. When the two switches are controlled by a signal in phase opposition, the resistance changes are mutually compensating and thus an *equivalent resistance* is obtained which is practically constant, Fig. 152.

Another advantage resides in the possibility of switching analog signals whose value can range from a few millivolts practically to the power supply voltages. Inverter circuits which can generate the control signals for analog switches can easily be made using CMOS technology (Fig. 153). Homogeneous, low power dissipation devices can be obtained since the inverter stage only dissipates power at the instant of switching.

In CMOS circuits, the substrates of *p* channel MOSs must be connected to the most positive potential and those of the *n* channel MOSs to the most negative

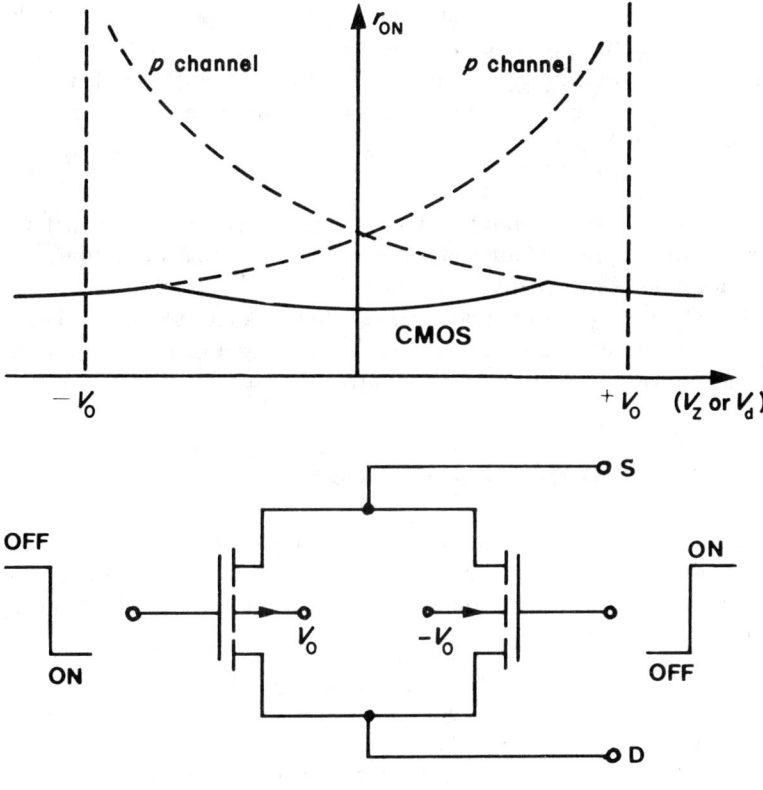

Figure 152

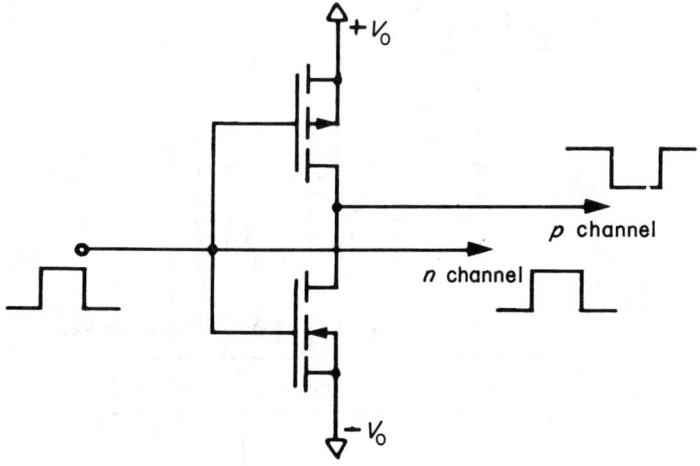

Figure 153

potential. This can result in what is termed the *parasitic thyristor phenomenon* because the structure of a CMOS is the same as that of a thyristor (the phenomenon is also known as 'latch up'). This can be avoided if a p^+ doped layer is inserted during fabrication between the substrate of the p channel MOS and the n channel MOS.

The advantages of CMOS devices used for switching are as follows:

in the ON state: low dynamic resistance (determined by the geometry and the values of the supplies), no 'Beta' effect, bi-directional operation, and small resistance changes with applied voltage.

in the OFF state: very high impedance, very low leakage currents and a switching speed comparable to that of other devices having a control circuit, since the speed is determined largely by the control circuit.

5.3.6 Examples of Actual Devices

Three examples of commercially available integrated switches will now be described, demonstrating in each case how the manufacturers solved the problems arising from the offset voltage due to saturation and from the dynamic resistance.

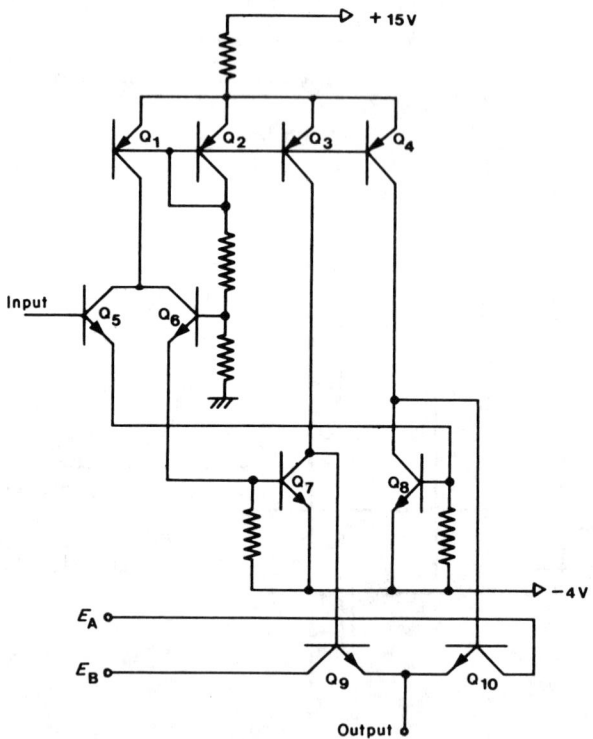

Figure 154

The first example is switch AD 555 from Analog Devices (Fig. 154).[60] It is a *quadrupole voltage switch* using bipolar transistors. The output is suitable for connection to two reference voltages E_A or E_B. The monolithic fabrication adopted by Analog Devices for this switch does not use *p-n* junction insulation but *dielectric insulation* (the semiconductor components are insulated from each other by a silicon oxide barrier). In this way low collector resistances are obtained for very small dimensions and hence the eight switches, together with their logic circuitry, can be on one chip.

Transistors Q_1 to Q_4 make up a triple current source and Q_5 and Q_6 constitute a comparator which feeds Q_7 or Q_8 according to the state of the control signal. When Q_7 is conducting (it takes the current supplied by Q_3), Q_8 and Q_9 are cut off. Therefore Q_{10} is saturated by the current from Q_4 and the analog input E is connected to the output.

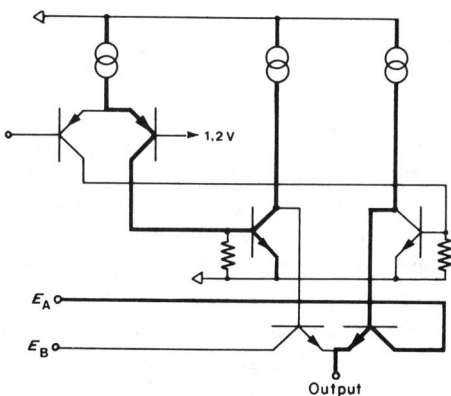

Figure 155

Figure 155 summarizes the operation of this switch when the control signal is in the state 1. The heavy lines indicate the transistors that are conducting and the current paths. This switch accepts voltages of up to 4 V on inputs A and B and has a response time of 5 μs. The offset voltage (saturation) due to transistors Q_9 or Q_{10} can be less than 2 mV and the resistances presented by the switches can be smaller than 25 Ω with a difference between the resistances presented by two switches of less than 10 Ω. Moreover, it is possible for the resistance values of the associated network to allow for the resistances presented by the switches.

The second example is a *current switch* using CMOS techniques.[27] The switch, an SPDT type, is used by Analog Devices in their DAC type AD 7520. Figure 156 shows the circuit diagram of one switch. This is a slightly different approach to the one described above because the output pair Q_8 and Q_9 are two *n* channel transistors instead of two complementary transistors. Transistors Q_4 to Q_7 constitute two inverters for controlling the output transistors in opposition. The geometry of transistor Q_1 and Q_2 is designed so that the input threshold is 1.4 V which makes the circuit compatible with DTL, TTL and CMOS logics. This system has no saturation or offset voltage and the equivalent resistances of the

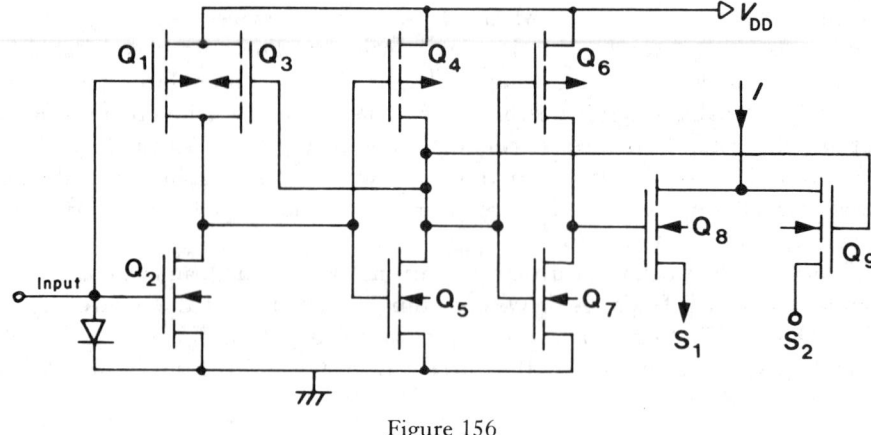

Figure 156

output transistors can be chosen by adjusting their geometry. However, it is difficult to make them negligible with respect to the resistors making up the associated network because this would increase the size of the device appreciably and hence its cost, and it is usually preferred to know their value accurately. Thus the switch associated with the bit of greatest weight will present a resistance of $20 \, \Omega$ and the switch corresponding to the bit of next least weight will have a resistance of $40 \, \Omega$, the next a resistance of $80 \, \Omega$, etc. The currents flowing in them will be in a geometric progression of common ratio one half. They will produce a constant voltage drop whose effect can be eliminated. The converter using these switches has a conversion time of 500 ns. Other manufacturers produce CMOS switches but these are usually of the SPST type, and Siliconix (DG 100, DG 200) and Intersil may be mentioned among others.

Bipolar current switches will be the object of the third example.[61] This system is used extensively and figures in the catalogues of many manufacturers, among them Analog Devices (AD 561, AD 562) Intersil (IH 8018), Fairchild, and Siliconix. These current switches are specially designed for use in weighted resistor DACs. A device of this type has already been described in the chapter devoted to DACs and its basic circuit diagram is given in Fig. 157. It has in one assembly, four current sources (transistors Q_1 to Q_4) and their associated switches (transistors Q_5 to Q_8). The devices vary according to manufacturer; in particular in the way circuits L_1 to L_4 are made up. The purpose of these circuits is to tie the voltage of the base of transistors Q_1 to Q_4 so that they are conducting or cut off depending upon the value of the control signal (state 0 or 1). Thus when the bit of greatest weight b_1 is in the state 1, the input diode D_1 is cut off, transistor Q_5 conducts and forces the potential of the emitter of Q_1 to become greater than that of the base and hence Q_1 is cut off. When b_1 is in the state 0 diode D_1 conducts. Because of circuit L_1, transistor Q_5 is cut off and transistor Q_1 conducts normally. If the associated network of resistors consists of weighted resistors then the currents through transistors Q_1 to Q_4 must be in a geometric progression. In order that the base to emitter voltages of these transistors remain equal, the emitter current densities must themselves be equal and in order to achieve this the areas of the base–emitter

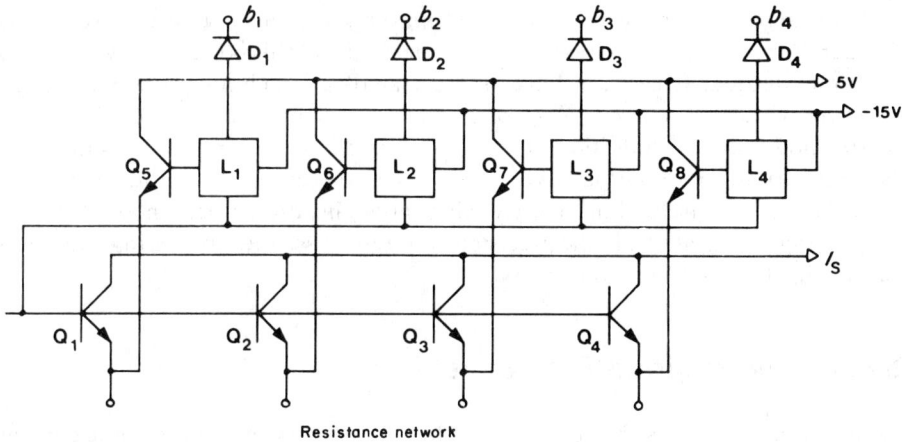

Figure 157

junction are in a geometric progression of the same common ratio as that of the currents. This technique is well suited to integrated fabrication and this explains the variety of circuits available. These are the fastest switches manufactured and make it possible to produce 12 bit DACs having a conversion time of 100 ns. The resistance presented by these switches is no longer of importance since we are concerned only with the output current. For reasons of space and overall size these circuits are limited to groups of four sources and four switches if weighted resistors are used (as in Fig. 157) or the groups can consist of ten switches and ten sources with ladder networks. This technique gives integrated circuits whose accuracy is compatible with 12 bit DACs.

5.4 REFERENCE SOURCES

The fourth important component used in the designs of DACs or ADCs is the reference source.[54] For instance in a DAC it is the reference source that provides the weighted currents which will then be summed if the corresponding bits in the word to be converted are equal to one. The stability and value of the signal affect directly the converter's performance. Let us consider a ten bit DAC having a reference source of 10 V. The sum of the inaccuracies due to the changes in temperature must be less than half a quantum, that is 4.88 mV. If the system must operate in a temperature range of 0–70 °C (commercial range) this demands a relative stability of supply of:

$$\frac{\Delta V_{\text{ref}}}{V_{\text{ref}}} = \frac{4.88 \times 10^{-3}}{10} \times \frac{1}{70} = 7 \times 10^{-6} \,^{\circ}\text{C}^{-1}$$

Therefore one must select a source with a stability better than 10 ppm °C^{-1}.

There are at present two ways of obtaining a high precision reference source. A temperature compensated *Zener diode* can be used and this is the most widespread solution adopted. It can also have an associated active circuit for adjusting the reference voltage or current. A second category of sources make use of *forward-biased* diodes and the addition of an amplifier will give the required voltage. At present very few monolithic DACs have an i.c. reference source and an external source must be used. This clearly highlights the difficulties encountered in fabricating integrated, high precision reference sources matching the performance of 12 or 14 bit converters.

5.4.1   Zener Diode Reference

When a diode is reverse biased there is a bias voltage for which the diode current suddenly increases (the avalanche effect). Thus a voltage source is created whose current practically depends only on the load connected across the diode terminals. Different values of this voltage, called the Zener voltage V_Z, can be obtained depending on the doping of the p and n semiconductors of the junction. However the voltage V_Z varies with temperature and the sign of this change depends on the value of V_Z (it is negative for $V_Z < 5.5$ V and positive above that). For a voltage V_Z of the order of 5.5 V the temperature has practically no influence and a stability of about 5 ppm °C^{-1} can then be obtained.

In order to take advantage of this very good stability it is necessary to use diodes having a Zener voltage of the order of 5.5 V and this greatly restricts the field of application. A first arrangement giving good stability for the voltage required for a given application consists in *compensating the diode for temperature changes*. This is done by connecting one or more forward-biased diodes in series with a positive temperature coefficient Zener diode, which hence have negative temperature coefficients (Fig. 158). For a given temperature range it is thus possible to compensate the Zener diode almost perfectly and obtain a stability of a few ppm °C^{-1} and the required voltage.

A second solution consists in adding to the Zener diode an *active circuit* so that a reference voltage or current is obtained having the required value and with the

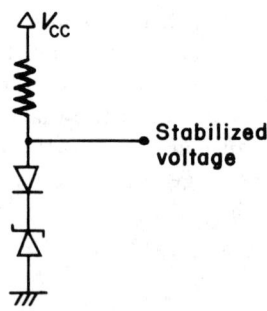

Figure 158

same stability as that of the Zener diode. This arrangement has several advantages; among which are that the diode can be supplied with constant current, a voltage obtained from a low impedance source can be available (capable of a certain output), and the generation of a current or voltage source is possible, etc.

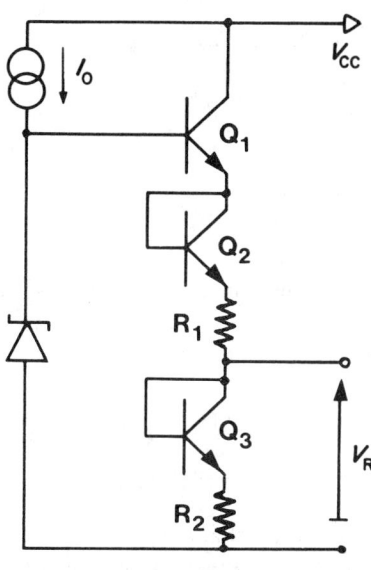

Figure 159

Figure 159 indicates a simple way of making such a device. The Zener diode is biased by the current source I_0 and a temperature dependent voltage V_Z is obtained at its terminals. This temperature dependence can be compensated by the voltages V_{BE} of three transistors (Q_2 and Q_3 are connected as diodes). Assuming these to be identical transistors and at the same temperature (which is possible for a monolithic circuit) then voltage V_R is given by:

$$V_R = \frac{R_2 V_Z + (R_1 + 2R_2) V_{BE}}{R_1 + R_2}$$

The variation of V_Z and V_{BE} with temperature are of opposite sign. By carefully selecting the components to be used we can make:

$$\frac{R_1 - 2R_2}{R_2} = \frac{\delta V_Z / \delta T}{\delta V_{BE} / \delta T}$$

and in this case the voltage V_R is independent of the temperature (at least as a first order approximation). Thus a stability of 30 ppm °C^{-1} can be obtained.

Another solution consists in using an operational amplifier with *two feed-back loops*, one positive and the other negative. Figure 160 shows the basic circuit diagram. This circuit can be considered as having two outputs; the voltage at the terminals of the Zener diode can be used, but this assumes that the load remains

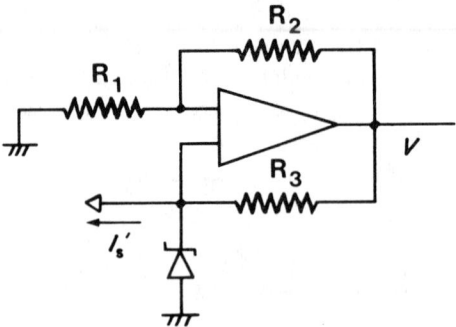

Figure 160

constant (current I'_S must be constant). If the load is likely to vary the output voltage of the amplifier can be used as a source. Assuming an ideal amplifier then the equations for calculating the available voltages and currents are:

$$V_S = \left(1 + \frac{R_2}{R_1}\right) V_Z$$

$$V_S - V_Z = R_3(I_Z - I'_S)$$

I_Z being the Zener current. In order that currents I_Z and I'_S can exist, the voltage V_S must be greater than the Zener voltage. The value of voltage V_S can be set by adjusting the values of resistances R_1 and R_2. If the amplifier is not perfect, in particular if it has offset currents and voltages, the first equation is modified and becomes:

$$V_S = \left(1 + \frac{R_2}{R_1}\right) V_Z + V_{OFF} + R_2 I_{OFF}$$

V_{OFF} being the input offset voltage of the amplifier and I_{OFF} the difference between the biasing input currents. Moreover this equation will give the voltage drift when the changes with temperature of the amplifier parameters are used for the calculation.

A current is often preferred to a voltage for use as reference in DACs. Since this current must be very stable it is obtained from a Zener diode as shown in Fig. 161. The input is at the reference potential and current I_{ref} can be set to the required value by adjusting the value of the resistance R_{ref} since:

$$I_{ref} = \frac{V_{ref}}{R_{ref}}$$

This current is available in the transistor emitter, the emitter voltage being determined by the circuit in which it is used (in the case of Fig. 161 this voltage can only be negative if the transistor is to conduct).

Figure 162 gives a circuit in use in which the Zener diode has a voltage of 6.2 V and the current I_{ref} is usually set at $\frac{1}{8}$ mA. The 2.2 kΩ resistor at the output of

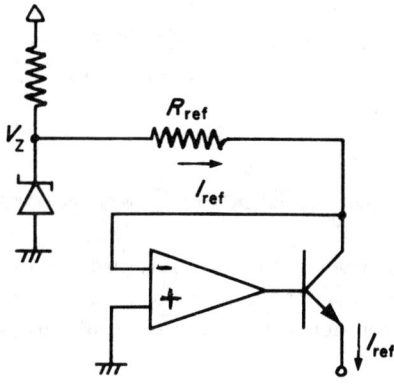

Figure 161

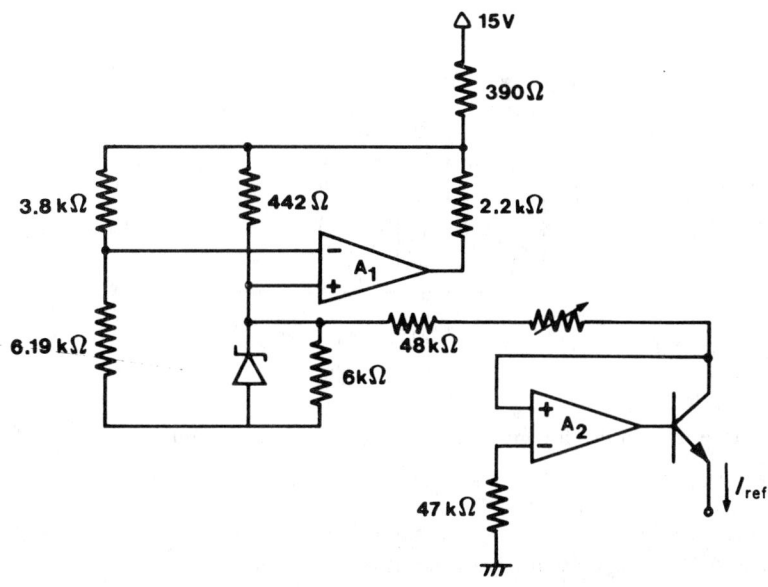

Figure 162

amplifier A_1 maintains the amplifier in the linear operating condition by reducing its loading, and the resistor also prevents the amplifier cutting off. The variable resistor allows the current to be adjusted to the value required. The 47 kΩ resistor connected to the output of the amplifier A_2 is used for compensating the effect of the offset currents of the amplifier.

However, Zener diodes have a drawback because they are an *appreciable noise source*. The effective value of this noise is related to the biasing current of the diode. This noise alters the stability of the reference voltage obtained and limits the use of the Zener diode as a reference standard.

Moreover the characteristics of Zener diodes vary with time and this results in the reference voltage drifting in a way which is difficult to estimate. The drift can be minimized by artificially ageing the components and it is then possible to achieve stabilities of 10 ppm $°C^{-1}$.

5.4.2 Reference Using Bipolar Transistors

Instead of using a Zener diode to obtain a reference source it is possible to make use of the voltage drop appearing across the terminals of a *forward-biased p-n junction*. This voltage drop depends upon the temperature and can be stabilized by using the circuit of Fig. 163.

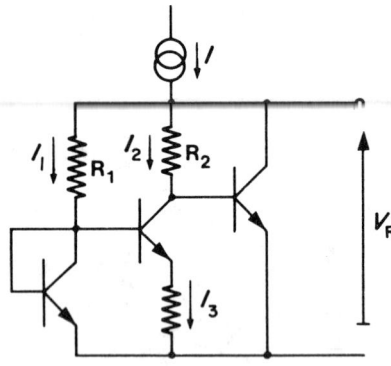

Figure 163

The current through a bipolar transistor obeys the equation:

$$I_E = I_S(e^{(q(V_{BE}/kT))} - 1)$$

I_S being the saturation current. Assuming that transistors Q_1 and Q_2 are identical their base to emitter voltages are related by:

$$\Delta V_{BE} = V_{BE1} - V_{BE2} = \frac{kT}{q} \ln \frac{I_1}{I_2}$$

This difference is positive and current I_1 is greater than current I_2. The voltage drop across resistance R_3 can be written as:

$$\Delta V_{BE} = R_3 I_2$$

The voltage given by the circuit is

$$V_R = V_{BE3} + R_2 I_2$$

$$V_R = V_{BE3} + \frac{R_2}{R_3} \frac{kT}{q} \ln \frac{I_1}{I_2}$$

The variation with temperature of these two terms is of opposite sign. By carefully selecting the resistances R_2 and R_3 and the currents I_4 and I_3 it is possible to eliminate in the expression for V_R the first order terms due to temperature variations.

This circuit will give a reference of low value (of the order of 1.2 V) which is only slightly temperature dependent, easily reproducible and having a lower noise level than a circuit using a Zener diode.

BIBLIOGRAPHY

1. K. S. Lion, *Instrumentation in Scientific Research*. McGraw–Hill Book Co., New York (1959).
2. H. Schmid, *Electronic Analog/Digital Conversions*. Van Nostrand, Reinhold, New York (1970).
3. D. H. Sheingold, *Analog/Digital Conversion Handbook*. Analog Devices Inc., Norwood (1973).
4. C. Shannon, *The Philosophy of Pulse Code Modulation*, pp. 1328–1331. Pire (1948).
5. J. A. Betts, *Signal Processing, Modulation and Noise*. English University Press, London (1970).
6. J. Dupraz, *Théorie de la Communication*. Eyrolles, Paris (1973).
7. P. J. Panter, *Modulation, Noise and Spectral Analysis*. McGraw–Hill, New York (1965).
8. J. Max, *Méthodes et Technques de Traitement du Signal*. Masson, Paris (1972).
9. A. Papoulis, Error analysis in sampling theory. *Proc. IEEE* **54**, 947 (1966).
10. A. Hellion, B. Escudie and P. Guillard, Influence de l'échantillonnage sur le traitement en temps différé des signaux rapides. Seminar GUTS–URSI–SEE, Grenoble (April 1974).
11. Y. Sevely, *Systèmes et Asservissements Linéaires Échantillonnés*. Dunod Université, Paris (1968).
12. B. M. Gordon, Digital sampling and recovery of analog signals, *EEE—Magazine of Circuit Design Engineering* (May 1970).
13. K. Cattermole, *Principles of Pulse Code Modulation*. Iliffe, London (1969).
14. H. Caurant, La modulation par impulsions et codage (MIC), *Toute l'Électronique* (June and August 1973).
15. J. Marcus, *Échantillonnage et Quantification*. Gautheirs–Villars, Paris (1965).
16. A. Sanchez, Understanding sample-hold modules, *Analog Dialogue* **5**, No. 4 (1971).
17. D. F. Hoeschele, *AD/DA Conversion Technique*. John Wiley, New York (1968).
18. R. Delsol, *Circuits Intégrés et Techniques Numériques*. Cepadues, Toulouse (1974).
19. D. Sheingold, *Analog/Digital Conversion Handbook*. Analog Devices Inc., Norwood (1972).
20. D. B. Bruck, *Data Conversion Handbook*. Hybrid Systems Corporation, Billerica, Massachusetts (1972).
21. D. Pinkowitz, When D/A converter glitches rear their heads . . ., *Electron Des.* **21**, 100 (25 October 1973).
22. C. R. Teeple, Don't forget DA converter tempo, *Electron. Des.* **22**, 110 (10 May 1974).
23. F. Dattée, État actuel de la théorie et des techniques d'application des amplificateurs opérationnels. Monograph of the review, *Mésures* (May 1972 to January 1973).
24. R. Demrow, Settling time of operational amplifiers, *Analog Dialogue* **4**, No. 1 (June 1970).
25. D. I. Dooley, A complete monolithic 10 bits D/A converter, *IEEE J. Solid-State Circuits* **8**, 404 (1973).
26. G. Kelson, A monolithic 10 bits digital-to-analog converter using ion implantation, *IEEE J. Solid-State Circuits* **8**, 396 (1973).
27. J. Cecil, A 10 bit monolithic CMOS D/A converter, *Analog dialogue* **8**(1), 3 (1974).
28. J. J. Hirsch, *Contribution à l'Étude des Conversions A–N et N–A au Moyen d'une Représentation Stochastique de l'Information*. Thesis for Engineering Doctorate, Université de Grenoble (1970).
29. A. Berg, A/D and D/A converter testing, *Electron. Des.* **22**, 64 (1 April 1974).

30. W. D. Miller, Pick the right DAC, *Electron. Des.* **22**, 110 (10 May 1974).
31. R. Best, Théorie systematique des convertisseurs A/D. Comportement statistique et dynamique, *Journées d'Électronique*, École Polytechnique Fédérale de Lausanne (1973).
32. W. J. Pratt, Don't lean on A/D specs, *Electron. Des.* **22**, 80 (12 April 1974).
33. Ch. Burniaux, Les convertisseurs analogiques/numeriques, *Toute l'Électronique* **41**, 49 (1974).
34. J. Barnes, Improve single-slope A/D accuracy, *Electron. Des.* **21**, 58 (18 January 1973).
35. M. A. Hassan, A new approach to the design of amplitude to code converters, *IEEE Trans. Nucl. Sci.* **20**, 77 (1973).
36. R. C. Kime, The charge-balancing A/D Converter: an alternative to dual-slope integration, *Electronics* **44**, 97 (24 May 1973).
37. G. Grandbois, Improved linear processing packs A/D converter into two IC chips, *Electronics* **45**, 93 (27 June 1974).
38. R. Alvesten, Calculate with a V/F converter, *Electron. Des.* **22**, 130 (7 June 1974).
39. P. Kinlmann and I. Glendan, Low-power A/D converter and multiplexer, *IEEE Trans. Instrum. Meas.* **23**, 149 (1974).
40. J. Whitmore, A 10 bit monolithic CMOS A/D converter, *Analog Dialogue* **9**, No. 2 (1975).
41. H. Duluta, Boost A/D rates with staggered operation, *Electron. Des.* **23**, 51 (15 February 1975).
42. T. W. Henry, High speed D/A A/D techniques, *IEEE Trans. Nucl. Sci.* **20**, 52 (1973).
43. W. D. Miller, Coding A/D converters for sign and magnitude, *Electronics* **45**, 110 (19 September 1974).
44. R. J. Tarver, Graphs aid selection of A/D converters, *Electronics* **45** (21 February 1974).
45. B. M. Gordon, Effects of noise on analog to digital conversion accuracy, *Journées d'Électronique*, École Polytechnique Fédérale de Lausanne (1973).
46. B. M. Gordon, Speaks out on what's wrong with A/D converters specs, *EEE— Magazine of Circuit Design Engineering* (1970).
47. H. Koller, New criterion for testing A/D converters for statistical evaluation, *IEEE Trans. Instrum. Meas.* **22**, 214 (1973).
48. G. Demas, Experimental verification of the improvement of resolution when applying perturbation theory to a quantizer, *IEEE Trans. Ind. Electron. and Control Instrum.* **20**, 236 (1973).
49. S. Cantanaro and G. Pallottino, Logarithmic analog-to-digital converters: a survey, *IEEE Trans. Instrum. Meas.* **22**, 201 (1973).
50. R. Dobkin, Logarithmic converters, *IEEE Spectrum* **6**, 69 (1969).
51. J. Giboons and H. Horn, A circuit with logarithmic transfer response over 9 decades, *IEEE Trans. Circuit Theory* **11**, 378 (1964).
52. *Synchro Conversion Handbook.* Data Device Corporation, Chatsworth, California (1974).
53. *Applications of Synchro/Digital Conversion.* Memory Devices (1975).
54. A. Grebene, *Analog Integrated Circuit Design.* Van Nostrand, New York (1972).
55. M. Elphick, Focus on comparator ICs, *Electron. Des.* **20**, 52 (October 1972).
56. C. E. Woodward, A monolithic voltage comparator array for A/D converters, *IEEE J. Solid-State Circuits* **10**, 392 (1975).
57. L. Clegg, High-precision thin-film resistance networks, *Analog Dialogue* **8**, No. 1 (1974).
58. T. Parello, Comparing thin film for precision resistor networks, *Analog Dialogue* **8**, No. 2 (1974).
59. J. King, Thick or thin-film resistors? *Electron. Des.* **22**, 92 (16 August 1974).
60. M. Krabbe, AD 555 monolithic quad switch, *Analog Dialogue* **5**, No. 2 (1971).
61. E. Maddox, Current-steering chip upgrades performances of D/A converters, *Electronics* **44**, 125 (4 April 1974).

INDEX

Absolute encoder, 132–133
AC digital converter, (AIDC) 147
AC–DC conversion, 145–147
AC–digital conversion, 144–156
Accuracy, 37, 42, 44, 62, 71, 79, 95, 101, 104, 105, 119, 121, 162
Acquisition time, 27–29
Active circuit, 182
Adapter circuit, 105
Amplitude
    noise effects, 109, 110
    of output voltage, 34
Analog delay lines, 5
Analog–digital conversion *see* Analog to digital conversion
Analog–digital converters *see* Analog to digital converters
Analog logarithmic converter, 122
Analog sampling, 11
Analog signals, 1
Analog switches, 170–181
Analog systems, 7
Analog to digital conversion, 67–130
    application fields, 3
    basic principles, 1
    operations carried out during, 7–9
    present state of technology, 1
    present status, 2
    theoretical considerations, 9–17
Analog to digital converter
    analog type, 69, 77–92
    bipolar, 72, 105, 106
    characteristic parameters, 69–72
    characteristics, 11–21
    classification, 68, 69
    definition, 67
    dual slope, 106
    errors, 72–77
    ideal characteristic, 70
    linearity, 114–116
    logarithmic, 121–130

logic
    operation, 98–101
    organization, 98–101
    performance, 101, 102
logic type, 69, 92–102
monolithic, 165
monotonic, 76
noise effects, 106–113
noise factor, 108, 109
non-monotonic, 76
parallel–serial, 92–95, 102
post subtractive, 97
presubtractive, 97
quasi-logarithmic, 121
reference, 115
serial or (sequential), 97
serial–parallel, 102
statistical test methods, 116–119
successive approximation, 126
testing criteria, 113–119
testing procedure, 114
transfer characteristic, 69
unipolar, 105
very high speed, 94, 102–105
Analog voltage, 128
Angular position–digital converter, 149
Angular sensor, 131–135
Aperture time, 26
Arithmetical operations, 4, 5
Automatic testing, 4
Automatic weighing systems, 7
Avalanche effect, 182
Averaging by addition, 5, 6

Binary code, 16, 17, 72
Binary coded decimal code, 18
Binary combinations, 30
Bipolar code, 17, 20–22, 49, 106
Bipolar current switch, 180
Bipolar transistor switches, 172, 173

Bipolar transistors, 186, 187
Boltzmann constant, 109, 172
Buffer, 29
  memory, 85
Butterworth filter, 13

Capacitive charge redistribution converter,
    58
Capacitive charge transfer converter, 88–92
  performance, 92
Capacitors, 88–91
Change-over switch, 55, 171
Charge error, 26
Charge transfer, 88, 90
Charge variation, 26
CMOS
  switches, 176–178
  techniques, 53, 179
  technology, 2
Codes, 17–22
Coding system, 16
Communications, 5, 6
Commutator, 171
Comparators, 86, 92–95, 99, 101–103, 105,
    158–164
  characteristics, 161–164
  definition, 158, 159
  fast, 164
  LM311, 163, 164
  noise, 107, 108
    factor, 109
  operational amplifier, 159, 160
  overdriven, 163
  production technology, 164
  slow, 164
  types of, 160
Computers, 6
Continuous counter ramp converter, 84–85
Conversion frequency, 71, 81, 92, 95
Conversion processes, 7
Conversion time, 35, 36, 70, 71, 82, 120, 181
Conversion time measurement, 64, 65
Correlation, 5
Cos theta (θ)
  available in analog form, 141
  available in digital form, 140
  determining sign of, 150
  in binary form, 148
  in digital form, 149
  method of obtaining, 141–144

Cosine multiplier, 152
Counter ramp converter, 82–84, 129
Counters, 86
Current generator, 86
Current sources, 3
Current switch, 179

DC–AC conversion, 137, 138
Delay lines, 5, 6
Demodulation, 145–147
Demodulator, 145, 146, 153
Depletion device, 173
Dielectric absorption, 25, 27
Dielectric insulation, 179
Differential amplifier, 146
Differential linearity error, 40–42, 62, 63,
    76
Diffusion, 166, 167
Digital–AC converter (DAIC), 135–144
  characteristics, 136
  DC–AC conversion, 137, 138
  parallel DAC, 136, 137
  using transformer, 138, 139
Digital–analog conversion
  see Digital to analog conversion
Digital–analog converter
  see Digital to analog converter
Digital analog multiplier, 32
Digital communications, 7
Digital comparator, 61
Digital computers, 1
Digital control, 44
Digital logarithmic converter, 123–125
Digital numbers, 7
Digital ramp converter, 129
Digital signals, 1, 54
Digital–synchro converter, 139–144, 157
Digital systems, 7
Digital to analog conversion, 30–66
  application fields, 3
  basic principles, 1
  definition, 30–32
  operations carried out during, 7–9
  present state of technology, 1
  present status, 2
  theoretical considerations, 9–17
Digital to analog converter, 3, 30, 96, 97
  bipolar, 34, 49, 50
  characteristic measurement, 62–65
  characteristic parameters, 33–37

Digital to analog converter (*cont.*)
  characteristics, 65
  classic type, 32
  classifications, 32
  commercial parallel type, 51–53
  direct, 32, 44
  error sources, 37–43
  exponential, 125, 126, 128–130
  indirect, 32, 44, 58–62
  intermediate parameter, 58–60
  non-linear, 125, 126
  non-monotonic, 76
  parallel type, 44, 54, 136, 137
  performance criteria, 62
  precision, 115
  principles of operation, 44–62
  reference, 63
  serial, 55–58
  specifications, 65
  testing, 62
  unipolar, 34
  weighted resistor exponential, 127
Digital voltmeter, 4
Dirac delta function, 10
Disc encoder, 131–133
Discrete voltages, 30
Distribution function, 110
Double integration, 79
Dual slope converter, 79–82
Dual-in-line packages, 2
Dynamic non-linearity, 26

Electromechanical sensors, 144
Electronic Scott circuit, 135
Elementary level of quantization, 14
Encoding operation, 7–9, 17
Enhancement devices, 174
Equivalent resistance, 176
Error probability, 107, 110, 113
Exclusive OR gates, 19
Excursion, 27

Feedback resistor, 127
Feedforward, 27
Field effect transistor (FET), 26, 29, 136,
    137, 173–175
Film resistors, 168, 169
Flip-flops, 61, 80, 99, 101, 164
Floating point representation, 124
Forward bias, 182

Forward-biased *p–n* junction, 186
Four quadrant multiplier, 53
Fourier transform, 10–13
Function generator, 5

Gain, 114, 163
Gain error, 25, 39, 43, 72, 74
Gaussian distribution, 109
Glitch, 36
Graphic display, 6
Gray code, 18, 19, 133

Hold
  mode, 29
  operation, 7, 26, 27
Hybrid addition, 5

Ideal low-pass filter, 10, 13
Ideal sampling, 9
  device, 23, 24
Ideal transfer characteristic, 161
Ideal transfer function, 69, 70, 72
Incremental encoder, 131
Incremental logarithmic converter, 125
Independence, 64
Industrial applications, 6
Information processing systems, 7
Input interface, 54
Instrumentation, 3, 4
Insulated FET, 174
Integral linearity, 76
Integrated circuits, 85, 92, 102
Interpolation function, 13
Interpolator, 13
Ionic implantation, 167

Junction FET switches, 175
Junction gate FET, 173

Ladder converter, 47–49
Ladder network, 128, 129
Large-scale integration technology, 2
Latches, 54, 55
Least significant bit (LSB), 18, 37, 48, 49,
    55, 56, 61, 83, 132
Linear approximation by parts, 123
Linear interpolation, 14
Linear quantization, 17
Linearity, 118
  criterion, 117
  definition, 39

Linearity (cont.)
  error, 39, 72, 74–76, 116
  gain, 75
  measurement, 62–64
Logarithmic amplifier, 122
Logarithmic converter, 121–130
  analog, 122, 123
  digital, 123–125
  incremental, 125
Logic, 17
  circuits, 84, 85, 93, 160
  diagram, 94
  matrix, 125
Low-pass filter, 13, 37, 59, 61

Magnitude sign code, 50
Matrix display, 6
Mean square error, 16
Modified representation, 124
Monotonicity, 41, 62, 76, 115, 118
MOS
  devices, 174
  techniques, 53
  technology, 2
MOS(CMOS) devices, 53
Most significant bit (MSB), 18, 19, 37, 48,
    49, 50, 61, 82, 83, 95, 132
Multiple threshold converter, 92
Multiplication operator, 4, 104
Multipliers, 152, 153
Multiplying converter, 32, 53
Multistate phenomenon, 107

NAND circuits, 94
Natural binary code, 18
Negative logic, 17
Noise
  effects, 106–113
    amplitude, 109, 110
    comparator, 107, 108
    errors, 110–113
  factor, 112, 113
    ADC, 108
    comparator, 109
  rejection, 120, 121
    factor, 71
    sources, 185
    voltage, 109, 112, 113
Non-linear quantization, 17
Non-linearity, 76

Octant selector, 144, 150, 151
Offset binary code, 20, 22, 49, 106
Offset current, 161
Offset error, 25, 29, 38, 42, 72, 73
Offset voltage, 73, 113, 114, 137
One's complement code, 21
Operational amplifier, 29, 122, 159, 160,
    164
Oscillation, 163
Oscillogram, 64
Oscilloscopes, 5, 6, 63
Output register, 99
Overall noise factor, 109

p-n junction, 122
Parallel converter, 92
Parasitic signals, 106
Parasitic thyristor phenomenon, 178
Parseval theorem, 13
Phase sensitive demodulator, 145
Phase shift, 14
Polarity detection, 159
Positive logic, 17
Postsubtractive converter, 97
Potentiometer effect, 95
Presubtractive converter, 97
Probability density, 61
Probability of occurrence, 61
Propagation time, 163
Pulse width modulation, 59
Pulses, 58–60, 78, 82, 83, 87, 88

Quadrant selector, 143, 150, 152, 153, 156
Quadrupole voltage switch, 179
Quantization
  characteristic, 15
  error, 15, 67, 70, 73, 78, 124
  noise, 15, 16
  operation, 7–9, 14–17, 67, 158
Quantum, 15, 17, 31, 34, 35, 62

Random generator, 61
RC circuit discharge, 123
Read only memory (ROM), 3, 142, 143, 155
Recovery process, 9, 12–14
Reference current, 182
Reference DAC, 63
Reference quantity, 44
Reference sources, 181–187
  bipolar transistors, 186, 187
  Zener diode, 182–186

Reference voltage, 4, 35, 51, 53, 88, 92, 93,
    95, 97, 103, 105, 106, 136, 137, 140,
    146, 147, 182, 185
Reflected binary code, 18
Rejection factor, 81
Resistance networks, 93, 165–170
Resistors, 38, 108, 127, 128, 181
    comparison of methods, 169, 170
    film, 168, 169
    obtained by diffusion, 166, 167
    obtained by ionic implantation, 167
    performance of, 170
Resolution, 35, 52, 70, 94
Resolver, 135
Response time, 101, 162, 163

Sample and hold
    circuits, 22–29
        errors, 25–28
    method, 12
Sample mode, 29
Sampling operation, 7, 9–12
Sampling pulses, 9, 10
Sampling rate, 9
Sampling theorem, 10
Saturation current, 186
Saturation resistance, 172
Sawtooth generator, 79
Scale factor error, 74
Scott connected transformer, 134, 139
Semiconductor devices, 2
Sequential converter, 55
Sequential floating point converter, 124
Serial hold converter, 56
Serial–parallel converter, 92
Settling time, 101
Shannon's theorem, 10
Shannon–Rack converter, 58
Shift register, 31, 54, 99
Shifted binary code, 20
Sign and magnitude code, 20, 22
Signal analysis, 5, 6
Signal generators, 3
Signal-to-noise ratio, 16, 17
Siliconix switch, 176
Sin theta (θ)
    available in analog form, 141
    available in digital form, 140
    determining sign of, 150
    in binary form, 148

in digital form, 149
    method of obtaining, 141–144
Sine multiplier, 152
Single phase converter, 147, 148
Single pole, double throw switch, 171
Single pole, single throw switch, 171
Single-slope converter, 77–79
Sinusoidal law, 142
Sinusoidal signal, 135
Slew rate, 102
Solid state technology, 102
Speed, 162
Speed error, 155
Spikes, 37
Stability, 181, 185
Staircase curve, 15
Statistical error, 116
Statistical test methods, 116–119
Stepped curve, 15
Stepped interpolation, 14
Stochastic converter, 60–62
Stochastic representation, 60
Successive approximation converter, 92,
    95–98
Successive approximation system, 130
Sum of elementary voltages, 32, 33
Sum of weighted voltages, 32
Summing amplifier, 46
Superposition
    error, 40, 64
    theorem, 48
Supply sensitivity, 43
Surface resistance, 167
Switching time, 174
Symmetric binary code, 20
Synchro–digital converter, 148–156
    characteristics, 157
    cos θ in digital form, 149
    sin θ in digital form, 149
    successive approximation, 154, 155
    theta (θ) in digital form, 149–156
    using ROMs, 155, 156,
Synchro transmitter, 133

Telemetry systems, 7
Television scanning, 6
Temperature effects, 42, 43, 64, 77, 182
Thermal noise, 108
Theta (θ) in digital form, 149–156

Thick film resistors, 168
Thin film resistors, 168
Tracking converter, 151–154
Transfer characteristic, 161
Transfer function, 31, 37, 38, 44, 53, 68, 69
Transformation, 44
Transformers, 134
  DAIC using, 139, 139
  Scott connected, 139
Transient analysis, 5
Transient output conditions, 36
Transient peak, 174
Transient recorders, 5
Transistor transistor logic (TTL), 17
Transistors, 52, 172, 173, 179, 183, 186, 187
Trigonometrical operations, 4, 5
True floating point representation, 124
Two's complement code, 20, 21, 49
Two-sectioned converter, 82

Unipolar codes, 17–19

Variable gain amplifier, 4
Variable threshold converter, 102
Variance calculation, 118, 119
Visual display unit (VDU), 6
Visual displays, 6
Voltage drop, 186
Voltage sources, 3
Voltage to frequency converter, 85–88, 123
Voltage–time converter, 78

Weighing systems, 7, 44
Weighted resistor converter, 45–47
Weighted resistors, 127, 128, 181
Weighted voltage, 96
Weighting converter, 95
Window, 115

Zener current, 184
Zener diode, 182
  reference, 182–186
Zener voltage, 182, 184
Zero order interpolation, 14